THE PUBLIC
TRUST DOCTRINE
and the Management of
America's Coasts

The PUBLIC TRUST DOCTRINE

and the Management of America's Coasts

JACK H. ARCHER

DONALD L. CONNORS

KENNETH LAURENCE

SARAH CHAPIN COLUMBIA

ROBERT BOWEN

Foreword by Jan Stevens

University of Massachusetts Press

AMHERST

Copyright © 1994 by The University of Massachusetts Press
All rights reserved
Printed in the United States of America
ISBN 0–87023–898–1
LC 93–48759
Set in Adobe Minion by Keystone Typesetting, Inc.
Printed and bound by Thomson-Shore, Inc.

Library of Congress Cataloging-in-Publication Data
The Public trust doctrine and the management of America's coasts / by
    Jack H. Archer . . . [et al.] ; foreword by Jan Stevens.
        p.   cm.
    Includes index.
    ISBN 0–87023–898–1 (cloth : alk. paper)
    1. Coastal zone management—Law and legislation—United States.
2. Public use—United States.   I. Archer, Jack H., 1938–   .
KF5627.P83   1994
346.7304′6917—dc20
[347.30646917]   93-48759
                                    CIP

British Library Cataloguing in Publication data are available.

*This book is published with the support and cooperation of the
University of Massachusetts at Boston.*

Contents

Foreword

California's distinguished chief justice Roger Traynor once observed that "though the law habitually moves in slow motion, it occasionally takes one step backward or two steps forward of remarkable span."[1] Few common law doctrines have moved forward as remarkably as the public trust. With roots in antiquity, the public trust has been embraced by state courts as a principle that guarantees public access to navigable waters, severely limits alienation of property by profligate sovereigns, and provides a basis, independent of police power, for the limitation of uses to which property may be subjected.

This book ties together, as few works have, the principles of the public trust, its many applications, and the manner in which it can constitute a tool for coastal management. Its appearance is timely. Like the policeman in *Pirates of Penzance*, the coastal planner's lot is not a happy one. Responsible planning must pursue a narrow channel, mindful of submerged hazards caused by takings decisions, but propelled also by statutory and common law obligations and the increasing realization that haphazard growth has led to pollution, proliferation of conflicting uses, shoreline erosion, and a denial to the people of their historic coastal heritage.

The lack of proper standards and certainty can lead to equal frustration on the part of those seeking shorezone projects. If comprehensive planning and objective guidelines are lacking, they may have to run through a veritable pinball machine of state, local, and federal regulations.

The recent experience of the South Carolina Coastal Council before the United States Supreme Court is an object lesson in the need to reexamine the public trust doctrine and its historical roots. In its recent pronouncements in *Nollan v. California Coastal Commission*, 483 U.S. 825 (1987) and *Lucas v. South Carolina Coastal Council*, 112 S. Ct. 2886 (1992) the Court has given notice that efforts by government to regulate coastal property will be strictly scrutinized.

---

1. "Statutes Revolving in Common Law Orbits," *The Traynor Reader* 166 (Hastings Law J. 1987).

Legislative findings will be met with skepticism rather than deference, and determinations of the validity of regulation may rest more on whether the harm to be prevented falls within eighteenth-century definitions of nuisance than on the existence of a rational legislative purpose.

Nonetheless, the *Lucas* decision sensibly recognizes that there can be no taking of property interests if they did not exist in the first place. To define what it calls "the range of interests that qualify for protection as 'property,'" the Supreme Court will continue to rely on state law. The court, however, has displayed some ambivalence in applying principles of federalism in areas of property law. While publicly avowing the black letter rule that of all bodies of law, property law was peculiarly a matter of state concern, *Corvallis Sand & Gravel Co. v. Oregon*, 429 U.S. 363 (1977), it occasionally shows fits of nervousness over the possibility that a state court may, in enunciating these rules, exceed settled expectations. See *Hughes v. Washington*, 389 U.S. 290 (1967) (Stewart, J., concurring); *Robinson v. Ariyoshi*, 441 F.Supp. 559, aff'd 753 F.2d 1468 (9th Cir. 1985), vacated, 106 S.Ct. 3269.

In *Summa Corp. v. California* (1984), Justice Rehnquist reiterated his perception of the public trust easement as one of "substantial magnitude," leaving the landowner with control of "little more than a naked fee." His observation was unnecessary to the holding but must have contributed to his conclusion that whatever trust interests existed in Mexican land grants had not survived the federal confirmation process.

The important thing about *Lucas* in this context is that it reaffirms the principle that state law must be the yardstick against which takings are measured. And state decisional law takes on added importance because mere conclusory assertions by a legislature will not suffice against a taking claim unless they are justified by "background principles of nuisance and property law." Suspicious of state legislatures and protective of property rights, the plurality opinion calls, somewhat desperately, for "an *objectively reasonable application of relevant precedents*" (emphasis added) before state principles of nuisance or public trust can be applied to eliminate all economically beneficial uses of property.

What does all this do to state courts' recognition that in the latter half of the twentieth century, public rights in tidal lands are not limited to commerce, navigation, and fisheries, but include recreational and ecological preservation? A hint may be found in Justice Kennedy's concurring opinion. He would ask, even in total dimunition of value cases, whether the limit imposed on property use is contrary to "reasonable, investment-backed expectations," measured by "objective rules and customs that can be understood as reasonable by all parties involved." "The takings Clause does not require a static body of state property law; it protects private expectations to ensure private investment." Id. at 2903.

The public trust has come a long way from *Arnold v. Mundy,* where New Jersey's chancellor was called upon to adjudicate a dispute over the oyster beds of the Rariton River. That simple controversy, in which heretofore "the strongest usually prevailed," led to the invocation of Vattel and Blackstone, the Magna Carta and the common law, the law of nature, and the civil law to hold that tidewaters and submerged lands are held in public trust. Few doctrines meet so well our Supreme Court's call for established common law by which takings claims may be measured.

This work, however, does not invoke the trust slavishly as a common law cure-all for coastal ills. It recognizes the desirability of statutory implementation of trust goals and the need for prioritization of trust uses. But it also outlines comprehensively some of the turns taken by state court decisions, for example, the New Jersey courts' application to coastal uplands, and California's determination that trust values must be observed and harm avoided in licensing water diversions.

The uneasy relationship between federal and state sovereigns is dealt with at length, both in terms of possible co-trusteeship and application of state trust principles to federal actions under the Coastal Zone Management Act and similar congressional directives. More and more often, questions arise as to application of the trust to lands once within state sovereignty, then conveyed or condemned for federal uses. Does, as a federal court in Massachusetts held, the trust emerge again, phoenixlike, when federal ownership ends, or has federal ownership wiped the title clean? As the closure of coastal military bases accelerates, the application of state trust principles to these often prime lands will become increasingly important. No one dealing with these problems should be without this book.

Finally, this work provides useful advice in an age of sometime judicial skepticism over regulatory planning. It suggests that trust principles provide a valuable underpinning, but no more than that, for coastal planning. Successful coastal regulation will rest on coherent legislative and administrative goals and on their intelligent and understanding implementation.

Jan Stevens
Deputy Attorney General,
State of California

Preface

We have written this book with the hope that the public trust doctrine will become an even more powerful tool through which we can protect, preserve, and use our coastal lands, waters, and resources in a responsible manner. As we hope this book makes clear, establishing the linkage between the techniques of coastal management that have been created in the United States during the past two decades and the evolving principles of the common law public trust doctrine can work to the advantage of coastal managers, developers, and citizens alike.

We would like to acknowledge the support of many individuals, organizations, and agencies in our efforts to assemble and analyze the materials on which the book is based. First, we wish to thank the Office of Ocean and Coastal Resource Management of the National Oceanic and Atmospheric Administration for its financial assistance in preparing an initial study of the public trust doctrine. Next, we owe our thanks to Arthur J. Rocque, Jr., Assistant Commissioner, Connecticut Department of Environmental Protection, and to the Coastal States Organization for their assistance in carrying out this NOAA-funded study. Also, we would like to acknowledge the research efforts of the following individuals from the firm of Choate, Hall & Stewart (Boston): G. d'Andelot Belin, Patricia Krumholtz, Ann Rickard Jackowitz, Mary Kate Whalen, Anthony Lazzara, and Jan Reitsma; from the University of Massachusetts at Boston: Ames Colt and John McKay; and from the President's Office, University of Massachusetts: Terrance W. Stone, Esq. Finally, we wish to thank our colleagues for their patience in reviewing and commenting on earlier versions of sections of the book, and to recognize particularly the critical and helpful comments of Robert W. Knecht, University of Delaware, and Marc J. Hershman, University of Washington.

THE PUBLIC
TRUST DOCTRINE
and the Management of
America's Coasts

I

The Development of the Public Trust Doctrine

A. SOME RELEVANT RECENT HISTORY

In the late 1960s the public in the United States dramatically changed its attitudes toward our nation's environment and the uses to which our lands and resources should be put. The public developed an increased awareness of the relationship between the environment and human health and welfare, and became intensely concerned about the quality of its physical environment and the appropriate regulation of land and resource use. The result was a proliferation of state and federal environmental and land use legislation,[1] often piecemeal, uncoordinated, and sometimes even conflicting, yet reflecting this common and deepening concern about the American landscape as a shared heritage and limited resource. Eight decades after Frederick Jackson Turner announced that we no longer enjoyed the luxury of an unlimited frontier[2] and four decades after the United States Supreme Court warned property owners that they could no longer use their land in a manner contrary to the public interest,[3] our citizens and lawmakers finally joined in a societal appreciation of the scarcity and fragility of our natural resources.

The public's new environmental consciousness did not overlook its most unique and valuable asset: its coastal areas, both ocean and freshwater, including the shores, submerged lands, tidewaters, and coastal ocean. In the preindustrial, sparsely populated colonial and pre–Civil War eras, these coastal areas seemed of little or no inherent worth. They were exploited (drained, mined, filled, and built

---

1. The vastness and complexity of our federal and state environmental laws can perhaps be adequately appreciated by scanning such multivolume works as F. P. Grad, *Treatise on Environmental Law* (4 vols., 1989), or W. H. Rodgers, *Environmental Law* (3 vols., 1986). An abbreviated but helpful review is provided by the annual *Environmental Law Handbook* published by Government Institutes, Inc., of Rockville, Md.

2. Turner, "The Significance of the Frontier in American History," *Report of the American Historical Association* (1983).

3. *Village of Euclid v. Ambler Realty Co.*, 272 U.S. 365 (1926). In this case the Supreme Court upheld the validity of comprehensive zoning regulations in general as a legitimate exercise of the police power, over objections that they violated constitutional due process guarantees.

over) with little appreciation of their loss. By the 1960s, however, the realities of eight generations of intensive economic development and explosive population growth had shattered the notion that the environment could forgive or absorb all trespassers. As the Council on Environmental Quality observed in 1970, in its first annual report, our coastal areas:

> represent attractive waterfront acreage in particular demand by industrial and commercial concerns and home buyers. Relatively inexpensive to dredge, fill and bulkhead for building sites, shallow wetlands attract many industries which are not dependent on waterfront sites but which find an economic advantage in developing these low-priced lands. Too often local governments acquiesce, anticipating the increased tax revenues. Consequently, natural coastal areas are being nibbled away. The long-range economic and ecological costs are borne not just by the particular local community, but by the people of the state and the region, and no less by the rest of the Nation.[4]

The council also noted the complexity of the demands upon the tidelands:

> Competition for the use of the limited coastal zone is intense. Shipping activities are increasing, with larger vessels needing deeper channels. Mining and oil drilling in coastal zones continue to enlarge their influence over these waters. Industrial and residential developments compete to fill wetlands for building sites. Airport and highway construction follows and further directs growth patterns in the coastal zone. Recreation—from enjoyment of the surf and beaches to fishing, hunting, and pleasure boating—becomes more congested as available areas diminish. Since over 90 percent of U.S. fishing yields come from coastal waters, the dependence of the commercial fisheries industry on a stable estuarine system is obvious.[5]

Officials and citizens began to realize that indiscriminate use of coastal areas had caused a grave and pressing situation, which needed to be addressed in a comprehensive manner:

> The key to more effective use of our coastland is the introduction of a management system permitting conscious and informed choices among development alternatives, providing for proper planning and encouraging recognition of the long-term importance of maintaining the quality of this productive region in order to ensure both its enjoyment and the sound utilization of its resources. The benefits and the problems of achieving national management are apparent. The present Federal, State, and local machinery is inadequate. Something must be done.[6]

The need for coordinated planning and development of coastal resources and activities soon found concrete voice in the federal Coastal Zone Management Act of 1972 (CZMA)[7] and the numerous state coastal zone management programs

---

4. Council on Environmental Quality, *Environmental Quality: The First Annual Report of the Council on Environmental Quality* 177 (1970).

5. Id. at 176.

6. *Our Nation and the Sea,* Report of the Commission on Marine Science, Engineering and Resources (the Stratton Commission Report), 49, and generally ch. 3 (1969).

7. 16 U.S.C. sec. 1451–64 (1972).

instituted pursuant to the federal law's provisions.[8] The state's basic authority to regulate its coastal zone derives from the same source as is used to enact zoning laws: each state's general "police power." The police power consists of those prerogatives of sovereignty and legislative power which are necessary for the protection of the health, safety, and welfare of state citizens and which the states did not surrender to the federal government when the United States Constitution was adopted.

The police power is a broad and valuable basis for the exercise of state regulatory authority (whether by statute or by agency regulations) that will rarely be invalidated (if challenged in the courts) so long as it is rationally related to a legitimate state goal and does not unduly burden interstate commerce.[9] However, the police power is subject to several limitations as the sole source of coherent coastal management programs. First, because of its breadth, the police power is not tailored to the special problems of coastal areas and does not provide organizing principles for addressing those problems, particularly the inevitable conflicts between legitimate but competing coastal uses.[10] Second, the police power has traditionally been linked to the restriction or suppression of harmful or dangerous activities and has not, until relatively recently, been utilized effectively to create and support affirmative, proactive management programs. Finally, and most limiting of all, if a state's regulation based on its police power goes "too far" in restricting uses or activities, it may be invalidated under federal and state constitutional guarantees against the taking of private property, even for public use, without the payment of just compensation.[11]

An additional and powerful source of authority that can be used in conjunction with the police power for more effective management of coastal areas is found in the public trust doctrine. This ancient property law principle, which arose out of our English common law heritage, exists in every state in the Union, but has only relatively recently begun to be fully appreciated. The traditional principles of the public trust doctrine are: (1) all tidelands and lands under navigable waters were owned by the original thirteen states at the time of the American Revolution, as successors in sovereignty to the English Crown, and each subsequent state was endowed with similar ownership rights at the time of its admission into the Union; (2) the states own these lands subject to a "public trust" for the benefit of all their citizens with respect to certain rights of usage, particularly uses related to maritime commerce, navigation, and fishing; and (3)

8. See ch. 3.

9. E.g., *Southern Pacific Co. v. State of Arizona*, 325 U.S. 761 (1945). See ch. 2, secs. D and G.

10. See ch. 2, sec. B.

11. See ch. 2, sec. H for analysis of the public trust doctrine and the changing law of takings after the recent decision of the United States Supreme Court in *Lucas v. South Carolina Coastal Council*, 112 S.Ct. 2886, 120 L.Ed.2d 798.

all lawful grants of such lands by a state to private owners have been made subject to that trust and to the state's obligation to protect the public interest from any use that would substantially impair the trust. Moreover, any such conveyed lands must be used by their private owners so as to promote the public interest and so as not to interfere unduly with the public's several rights under the public trust doctrine.[12]

B. THE UNIQUE POTENTIAL OF THE PUBLIC TRUST DOCTRINE

The public trust doctrine provides states with a unique supplemental authority over their coastal lands to be used in conjunction with the traditional police power authority.[13] First, the historical origin of the public trust doctrine stems from concern for the particular benefits and uses of coastal areas. Therefore, the doctrine as it has been implemented in the course of its ancient and modern history provides a unique perspective on specific issues of importance in coastal land management. Second, because the public trust doctrine is fundamentally a property- or ownership-based doctrine, a state's authority under the public trust doctrine is not limited to the power to regulate but also includes the power to protect the state's fundamental rights in its property, and the rights of all members of the public to use such property, even when the property has been conveyed into private ownership. Third, because the public trust doctrine is grounded in property ownership principles, it is less vulnerable to a challenge by a private property owner based upon the takings clause of the United States Constitution when a state has exercised its rights and obligations as a trustee over public trust land to restrict (or even prohibit) the activities of private land-owners. Fourth, the public trust doctrine has its origins in the common law and in the basic premise that scarce coastal resources are to be used in a manner consistent with the public interest. Consequently, although the core of the public trust doctrine has remained stable over the past two centuries, it shares the inherent common law capacity to grow and adapt in response to the changing social conditions and public needs which has distinguished the development of Anglo-American legal doctrines.[14] Finally, the public trust doctrine has at its root

12. See below, ch. 1, sec. C and ch. 2, secs. B and F.

13. The increased application of the public trust doctrine by state coastal managers is particularly appropriate in light of Congress's 1972 declaration, in the federal Coastal Zone Management Act, that "the key to more effective protection and use of the land and water resources of the coastal zone is to encourage the states to exercise their full authority over the lands and waters in the coastal zone." 16 U.S.C. sec. 1451(i). In addition to the public trust doctrine and the police power, states can protect and advance the public interest in their coastal areas by utilizing their power to enjoin public nuisances (see below ch. 2, sec. E.) and their ultimate power to take private land by eminent domain (see below ch. 2, sec. H.).

14. See ch. 2, secs. A and B for examples of such legal development affecting the public trust

the notion of a legally enforceable "trust," which is analogous, in many senses, to the body of private and charitable trust law that has developed in the United States. This analogous law is a valuable source of insight and precedent for determining the rights and responsibilities of the states as trustees over public trust lands for the benefit of their citizens, particularly in jurisdictions where there has been little concrete legislative or judicial application of public trust principles.[15]

C. EVOLUTION OF THE PUBLIC TRUST DOCTRINE IN THE UNITED STATES

Despite arguments over its historical origins, development, and scope, there appears to be general agreement that the public trust doctrine existed as an important principle of English common law at the time of the English settlement of the New World in the early seventeenth century.[16] The central idea of the public trust doctrine has always been that each state holds its coastal waters and the land underneath them for the benefit of the public. Until recently, however, American coastal managers have tended to use the public trust doctrine primarily as a legal restriction rather than as an affirmative management tool.

The Massachusetts Bay Colony's Ordinances of 1641 and 1647, in which the doctrine made its American debut, foreshadowed its thrust in American law over the following three centuries. Under those ordinances, the colony granted title extending to the low watermark to owners of land adjoining tidally influenced waters and granted title to the high watermark to owners of land adjoining "great" ponds over ten acres in size. The purpose of the ordinances was to encourage private wharf building for the stimulation of commerce, which the

---

doctrine. In this connection, the fact that the public trust doctrine historically included maritime "commerce" as one of the quintessential public rights protected by it may provide the basis for additional common law development, at least as to water dependent and water-related activities, should state courts be inclined to interpret the term "commerce" as expansively as the federal courts have done under the interstate and foreign commerce clause of the United States Constitution, Article I, cl. 3, sec. 8. See, e.g., *Gibbons v. Ogden*, 9 Wheat. (22 U.S.) 1 (1824); *United States v. Appalachian Power Co.*, 311 U.S. 377 (1940); *United States v. Rands*, 389 U.S. 121 (1967).

15. See ch. 2, sec. C.

16. The various views and controversies regarding the origin, development, scope, and proper application of the doctrine, as well as references to virtually all the significant judicial decisions, books, and articles on the subject, can be found in 19 *Environmental Law* No. 3 (Spring 1989), an important Symposium on the Public Trust and the Waters of the American West: Yesterday, Today and Tomorrow, held under the auspices of Northwestern School of Law of Lewis and Clark College. As fascinating and informative as these discussions are, the debate about the origins of the public trust doctrine is largely of academic historical interest, because it has been so clearly acknowledged by all American courts and therefore exists as a binding legal principle. See, e.g., *Phillips Petroleum Co. v. Mississippi*, 484 U.S. 469, 477–78 (1988).

struggling colonial government could not afford to do. Despite this expansion of private ownership in apparent derogation of the traditional public trust rights, the ordinances expressly reserved for the public the right to use these areas for fishing, fowling, and navigation. In addition, the public retained the right to pass over the private land itself between the high and low watermarks or next to great ponds in order to exercise its reserved rights.

A century and a half later, with the Northwest Ordinance, the newly formed United States government adopted public trust principles to apply to its great internal waterways. As one of its first acts, the national government made the Mississippi and St. Lawrence rivers "common highways, and for ever free."[17] The Massachusetts Ordinances of 1641 and 1647 and the Northwest Ordinance significantly expanded the scope of the public trust doctrine beyond that recognized in England, to include fresh as well as tidal waters. This extension of the doctrine to fresh waters, adopted by most states as the country grew, was the first reflection of the doctrine's dynamic nature in the United States.

1. The Influence of English Common Law

Although the public trust doctrine initially appeared in the United States in statutory form, it has developed mainly through judicial decisions, mostly dealing with disputes over title or access to public trust lands. After the Revolution, American courts faced the daunting prospect of creating a new body of case law to replace the English common law. The courts and most state legislatures rejected this option and instead adopted, either by judicial decision, by "common law reception" acts, or by constitutional provision, the familiar English common law that had governed the colonies before the Revolution.[18] Both state and federal courts referred to pre-Revolutionary English common law as the direct ancestor of, and thus the presumptive authority and precedent for, application of the common law in the United States. The public trust doctrine's place in English common law accordingly gave it an automatic foothold in the common law of the new United States.

Under seventeenth- and eighteenth-century English common law, coastal lands were subject to two different ownership interests. The Crown held title to the coastal waters and lands beneath them by virtue of its sovereignty. This interest was called the *jus privatum*, which was a transferable property interest

---

17. The Northwest Ordinance of 1787, 1 Stat. 50 (1789); See also *Barney v. City of Keokuk*, 94 U.S. 324 (1876).

18. See, e.g., *Livingston v. Jefferson*, 1 Brock. 203, Fed. case No. 8, 411 (Cir. Ct. Va. 1811); *Fitch v. Brainerd*, 2 Day 163 (Sup. Ct. Errors, Ct. 1805); 1 Laws of the Commonwealth of Pennsylvania, c. 726 of the session ending Jan. 21, 1777; Mass. Const. of 1780, ch. 6, sec. 6.

held by the Crown or by private persons claiming through it.[19] The *jus privatum* was, however, subject to the public trust rights of the English people, who held the *"jus publicum"*—the common right to use these public trust lands and their resources for certain traditional purposes necessary to individual survival and livelihood, including navigation, commerce, and fishing.[20]

This division of ownership interests between the Crown and the people restricted the use of public trust land. The *jus privatum* in public trust land, whether held by the Crown or by an owner claiming through the Crown, theoretically yielded to the paramount *jus publicum* of the English people. Therefore, the holder of the *jus privatum* could not impede the public's customary use of public trust land for navigation, commerce, and fishing. But the common law's recognition of divided ownership interests in trust lands created a tension between private and public rights. Private ownership inevitably led to conveyances of public trust lands for purely private or non-water-dependent or -related purposes inconsistent with the common rights of usage. At the same time, public rights continually posed a threat to the individual's investment in what he perceived as his land. Neither the English common law nor its American descendants have established clear-cut mechanisms to resolve this tension. Thus modern coastal managers and advocates are faced with both difficulties and creative opportunities in their discretionary applications of public trust principles.

2. The Early Development of the Public Trust Doctrine in the United States

The reshaping of the public trust doctrine in the United States, despite the preservation of its basic outlines and core policy, has reflected the political and geographical differences between the United States and England. Recognition of these differences gave American courts the discretion to emphasize those facets of the doctrine they favored and the justification to ignore those they did not. Two political differences between England and the United States are primarily responsible for the doctrine's dynamic and somewhat uneven development in American law.

First, unlike England, the United States is a federal system with certain powers held by the individual states and others by the national government. The creation and regulation of property rights within their borders is one of the powers held by the individual states. Thus, in the United States, use of coastal lands and waters is primarily a matter of state law, although the state's authority over coastal waters yields to the federal government's supreme authority under the

---

19. Hale, *De Jure Maris,* in 1 A Collection of Tracts Relative the Law of England 84, 89 (F. Hargrave 1st ed. 1787).

20. Id.

Constitution to regulate navigation and commerce (often called the "federal navigational servitude"). In addition, new states entered the union over a period of more than 160 years, during which time the doctrine has undergone change (along with virtually every other body of law) in response to society's changing needs and demands. Thus, although at the time of its admission into the Union each new state possessed exactly the same rights and obligations as did the original thirteen, including those relating to public trust lands, the public trust doctrine has developed somewhat differently in each state.[21]

Second, in each of the states the people are the sole source of legitimate authority and they collectively exercise this authority through their state govern-- ments. The sovereignty that vested in the Crown in England now inheres in the people of each state. The common law formulation of public trust ownership interests necessarily undergoes a modification when inserted into this system. The *jus privatum* and *jus publicum* still make up the property interests in trust land, but now what is essentially the same entity—the people of the state and the state government as the representatives of those people—holds both interests.[22] The people of each state have empowered their popularly elected legislatures to enact laws and regulations governing public trust lands as well as all other matters of public interest. But, in our modern era, when there are too many complex problems requiring ongoing expert supervision and regulation for a legislature itself to handle, the legislatures of each state have generally delegated regulatory authority over numerous matters of public concern, including the management of coastal resources, to expert administrative agencies, which have become our society's most characteristic forum for governmental activity.[23]

One respected state court at an early date described the nature of the state's sovereign authority over public trust lands:

> The people . . . may make such disposition of [the beds of navigable waters], and such regulation concerning them, as they may think fit; that this power . . . must be exercised by them in their sovereign capacity; . . . that the legislature . . . may lawfully erect ports, harbours, basins, docks, and wharves; . . . that they may bank off those waters and reclaim the land upon the shores; that they may build dams, locks, and bridges for the improvements of the navigator and the ease of passage; that they may clear and improve fishing places. . . . The sovereign power itself . . . cannot, consis-

---

21. See, e.g., *Shively v. Bowlby,* 152 U.S. 1, 57 (1894).

22. Since a single person or entity cannot be both sole trustee and sole beneficiary of a trust [*Restatement (Second) of Trusts* 115(5) (1959)], it might at first glance be thought that trust principles are inapplicable, in the American system, to public trust analysis if the people are in effect both the trustee and the beneficiary. This is, however, a simplistic and inaccurate characterization of our democratic system. *See* ch. 2, sec. C for a more thorough discussion of the analogy to private trust law.

23. See ch. 2, sec. D, dealing with delegation of public trust responsibilities.

tently with the principles of the law of nature and the constitution of a well ordered society, make a direct and absolute grant of the waters of the state, divesting all the citizens of their common right.[24]

The United States Supreme Court has consistently held that this authority is initially the same in every state of the Union, under the "equal footing doctrine," which originated in the Northwest Ordinance of 1787.[25] The equal footing doctrine dictates that each new state, upon entering the Union, has the same sovereignty and jurisdiction over all territory within its limits as was enjoyed by the thirteen original states, which in turn had inherited all sovereign rights of the English Crown.[26] This sovereignty is defined not by the law of the nation that held the territory before its entry to the United States, but by the common law of England as modified in the United States.[27]

Under the equal footing doctrine, each state has complete power over its public trust lands, subject only to the federal government's supreme power under the United States Constitution's commerce, navigation, and treaty powers.[28] A state may, however, decide to limit the waters over which this power extends.[29] Thus all states begin with equal sovereign jurisdiction over all their public trust lands, but the individual states may decide to extinguish their rights or to uphold grants of certain public trust lands to private individuals.[30]

Each state holds its public trust lands "in trust for the enjoyment of certain public rights."[31] From an early date, courts have acknowledged that a state may regulate the methods by which the public enjoys its rights, to better preserve and

24. *Arnold v. Mundy*, 6 N.J.L. 1, 78 (1821). The Supreme Court of the United States in *Martin v. Wadell*, 14 U.S. 345, 349 (1842), echoed the *Mundy* view of state sovereignty and gave it the authority of the nation's highest court.

25. "And whenever any of the said states shall have sixty thousand free inhabitants therein, such state shall be admitted by its delegates into the Congress of the United States, *on an equal footing with the original states in all respects whatever.*" Northwest Ordinance, Maxwell's Code, XII. See also *Phillips Petroleum*, 484 U.S. at 472.

26. See *Escanaba Co. v. Chicago*, 107 U.S. 678, 689 (1883). The case of *Pollard's Lessee v. Hagan*, 44 U.S. 212 (1845), established the relationship between the equal footing doctrine and the public trust doctrine. In that case, the Court nullified a federal grant of intertidal Alabama land made prior to Alabama's statehood. The Court found that the equal footing doctrine restricted the United States to the role of caretaker of a future state's interests in the interim between foreign rule and statehood. The United States had no power to grant to a private person the soil under navigable waters and thus deprive the forthcoming state of its sovereign authority over such land. Id. at 230. Any federal grant was thus limited to the area above the high watermark. Id. at 219, 229.

27. Id.

28. See generally ch. 4, dealing with the federal role in the public trust doctrine, and, e.g., *Shively v. Bowlby*, 152 U.S. 1 (1894); *Knight v. U.S. Ass'n*, 142 U.S. 161 (1891).

29. *Hardin v. Jordan*, 140 U.S. 371, 382 (1891).

30. Id. at 383. See *Phillips Petroleum*, 484 at 483; and ch. 2, sec. F dealing with conveyances of public trust lands and termination of the trust.

31. *Smith v. Maryland*, 59 U.S. 71, 74 (1855).

promote that enjoyment.[32] As examples, states may forbid certain methods of oystering in order to protect the growth of oysters,[33] grant exclusive fisheries,[34] and license private wharves and levees in order to encourage navigational uses.[35] The breadth of the state's authority to regulate public trust lands is directly related to the public interests that the doctrine is intended to protect. Thus courts have approved some state conveyances of public trust land to private parties for purposes consistent with the public trust, subject to certain standards established by the Supreme Court of the United States and the various state courts prescribing the conditions for such conveyances.[36]

3. The Definitive Supreme Court Cases

Four United States Supreme Court cases dominate our modern understanding of the public trust doctrine. They deserve brief description and should be known by coastal managers, public officials, and others interested in coastal management, because state and federal courts frequently refer to them in decisions dealing with public trust rights and because they will be referred to frequently in the remainder of this book.

Two of these cases occurred in the 1890s. In the first, *Shively v. Bowlby,* the Court undertook a full review of the doctrine's judicial past in order to resolve the "diversity of view as to the scope and effect of the previous decisions of this court upon the subject of public and private rights in lands below high water mark of navigable waters."[37] The Court started with the then prevalent view that the public trust doctrine naturally arises out of the tidelands' "uselessness" for private occupation, cultivation, or improvement.[38] These lands' natural and primary uses are the public ones of navigation, commerce, and fishing.[39] Thus the doctrine's purpose is to ensure that these lands are put to their highest and best uses in the public interest.[40]

32. Id. at 75.

33. Id.

34. *Hardin,* 140 U.S. at 382.

35. *Barney v. Keokuk,* 94 U.S. 324, 342 (1877).

36. See ch. 2, sec. F, dealing with conveyances of public trust lands and termination of the trust.

37. *Shively,* 152 U.S. at 10–11. The Supreme Court in *Shively* noted that "there is no universal and uniform law on the subject [of the public trust doctrine]. . . . [E]ach state applies the doctrine . . . according to its own views of justice and policy"; and the court provided the still timely warning that, despite a fundamental core of public trust principles shared by all coastal states, "[g]reat caution . . . is necessary in applying [public trust] precedents in one state to cases in another." 152 U.S. at 26.

38. Id. at 11.

39. Id.

40. After a summary of the tenets of the doctrine developed by earlier decisions of the Supreme Court, the Court in *Shively* held that the federal government can make grants of public trust land to private parties in a territory. This power goes beyond the earlier announced authority to "honor foreign grants" recognized in *Knight,* 142 U.S. 161. The *Shively* Court held that the federal govern-

The second key Supreme Court case of the 1890s, perhaps the most celebrated public trust decision in American history, involved a state grant of virtually the entire Chicago waterfront to a private party. In 1869 the Illinois legislature had granted to the Illinois Central Railroad "all the right and title of the State of Illinois in and to the submerged lands constituting the bed of Lake Michigan"[41] for a certain distance, in consideration of 7 percent of the railroad's gross income from its use of these lands. The grant required that the railroad hold the fee forever and prohibited the railroad from obstructing the Chicago harbor or impairing the public right of navigation, but contained no other restrictions on the property's use. A few years later, a differently constituted Illinois legislature revoked the overly generous grant, and the railroad challenged the validity of the revocation in court and ultimately brought the challenge before the United States Supreme Court.

The Supreme Court upheld the legislature's power to revoke its earlier grant, explaining that a state may, in certain circumstances, grant parcels of submerged lands to private individuals for building wharves and docks, because these grants further the public's interest in navigation and commerce. A state may not, however, convey away the submerged lands of an entire harbor, bay, or lake.[42] The critical issue was not just the quantity of land conveyed, but the loss of a state's ability to control an entire waterfront area. The Court stated as a general proposition that a state's conveyance of any piece of public trust property which impairs the public's interest in remaining public trust lands is void.

The unusual facts of *Illinois Central,* where almost the entire Chicago waterfront on Lake Michigan lay in the control of the railroad, obviously influenced this ruling. The Court's statement that "[t]here can be no irrepealable contract in a conveyance of property by a grantor in disregard of a public trust, under which he was bound to hold and manage it,"[43] would literally void any coastal conveyance not serving a public trust purpose. But a cogent argument can certainly be made that such a sweeping prohibition, appropriate in the extreme case where "the harbor of a great city and its commerce have been allowed to pass into the control of [a] private corporation,"[44] may not be necessary when dealing with the more common situation where a state conveys much smaller individual parcels primarily to further public interests.

---

ment can grant such lands for any public purposes appropriate to the objects for which the United States holds the territory. 152 U.S. 57–58. But because the United States had never done this other than to honor international treaties over new territories, Congress must clearly express its intention to break with past practice when purporting to convey land below the high watermark. Id.

41. *Illinois Cent. R.R. v. Illinois,* 146 U.S. 387, 406 n.1 (1892).

42. Id. at 453.

43. Id. at 460.

44. Id. at 455.

The Supreme Court recognized this distinction less than thirty years later when it held that in certain circumstances a state can terminate its control over public trust land by conveying it to private parties. *Appleby v. City of New York*[45] involved a conveyance by New York City, with the state's approval, of a relatively small portion of land submerged under the Hudson River. When the city later tried to regulate and preserve navigation over this area, the Court sustained the owner's trespass action against the city. According to *Appleby*, as long as the state's statutory grant was made for money consideration and in support of a definable public purpose, the state no longer holds the *jus privatum* or the *jus publicum* and has no legal right to use the water over the conveyed land.[46] The ground rules for state conveyances of public trust lands must be found somewhere between *Illinois Central* and *Appleby*.[47]

The public trust doctrine lay relatively dormant in the sixty years following *Appleby*. But new battles over coastal areas emerged with the growing concern over development and the push for conservation of natural resources in the 1960s and 1970s. Following a plethora of state cases, the Supreme Court revisited and significantly extended the reach of the public trust doctrine in 1988. In *Phillips Petroleum Co. v. Mississippi*,[48] the Court held that the doctrine extended in every state to all tidal waters, regardless of their navigability. The Court employed a broad interpretation of state public trust authority to reconcile its decision with evidence that the original states did not claim title to non-navigable tidal waters. According to the Court, every state upon its admission to the Union received ownership of all lands under waters subject to the ebb and flow of the tide, as well as to lands beneath navigable fresh waters. As a result, the states could "define the limits of the lands held in public trust and recognize private rights in such lands as they saw fit."[49]

The *Phillips Petroleum* decision may have significant implications for future exercises of state authority over public trust lands. Each state entered the Union with the ownership of all tidal and navigable waters and the lands beneath them and, apparently, can narrow its responsibilities as it "sees fit." In addition, a state can presumably expand its public trust authority over these lands and waters. In rejecting navigability as the sole benchmark of public trust applicability, the Supreme Court pointed to a "[s]tate's public trust interest in these lands . . . such as bathing, swimming, recreation . . . and mineral development."[50] The

45. 271 U.S. 364 (1926).
46. Id. at 397–99, 402–403.
47. See ch. 2, sec. F for a discussion and summary of these rules.
48. *Phillips Petroleum*, 484 U.S. 469.
49. Id. at 473–74 (citing *Shively*, 152 U.S. at 26).
50. Id. at 482.

Court also noted that lands beneath tidal waters had in some states been validly filled and conveyed "to create land for urban expansion."[51] Thus it is apparent that a state may increase the universe of public trust uses beyond the traditional areas of navigation, commerce, and fishing, as well as narrow its involvement by granting private rights in these lands. Every state starts with the same rights under the equal footing doctrine, but the precise interpretation of the doctrine is left to the judicial and public policy decisions of each individual state, so that the outcomes in the several states will necessarily differ.

D. THE PUBLIC TRUST DOCTRINE AS A TOOL FOR COASTAL MANAGEMENT

Despite two centuries of legal development, and despite its fundamental core principles, several key characteristics of the public trust doctrine remain in flux. Significant divergences about these issues exist among the coastal states, and these states vary widely in the extent to which their legislatures, judiciaries, and administrative agencies have anticipated, refined, and applied the public trust doctrine and its underlying principles. The remaining parts of this study are intended to provide individuals and organizations interested in coastal management (state coastal agencies and managers, state attorneys general, and other state officials, including state judges, as well as environmental advocates, developers, title companies, property owners, and other private citizens) with an understanding of and insight into critical aspects and applications of the public trust doctrine. The goal is not only their better understanding of the doctrine in their states, but more important, their enhanced ability to coordinate and balance the legitimate interests of development and preservation of coastal lands, waters, and resources. In light of the doctrine's different applications among the states, it may be helpful for readers to use the following basic checklist of issues to guide their analysis of the doctrine in each state:

What is the geographic reach of the public trust doctrine in the state, and how are the boundaries actually measured?

What are the appropriate or protected uses of public trust lands in the state, and how is the choice made between legitimate competing uses?

What state entity—the legislature, a state agency, or some other entity or official—is empowered to act as the trustee over public trust lands?

How has the public trust doctrine been enforced, and is there explicit or implicit authority for enforcement in any of the state statutes, regulations, or programs?

51. Id. at 476.

Does the state have the authority to convey public trust lands and terminate the trust, and if so, how is this done?

Does the state have authority to issue licenses and leases for public trust lands, and if so, what mechanism or procedure should be used to effect those licenses and leases?

What, if any, precautions are necessary to protect against challenges, under the taking clauses of the state and United States Constitutions, to the state's exercise of its public trust authority?

How does or could the public trust doctrine influence state decision making in the most commonly occurring or controversial areas of coastal management?

What role, if any, does the federal government play under the public trust doctrine?

Because the public trust doctrine is essentially a state law doctrine and therefore differs from state to state, this study cannot definitively answer the questions outlined above. However, it is structured to provide both the background information needed to conduct this analysis and ideas and possibilities for application of the public trust doctrine in reaching beyond its existing uses. The checklist of questions listed above is keyed to the subsequent portions of this study and is accordingly relatively general. A more detailed checklist is found in the appendix.

▌▌
Significant Issues in the Application of the Public Trust Doctrine

A. GEOGRAPHIC SCOPE OF THE PUBLIC TRUST DOCTRINE

This section addresses several threshold issues under the public trust doctrine. First, what lands are public trust lands? Although generally the states define their public trust boundaries in a similar fashion, some states do so differently. This question breaks into two parts: what is the state's definition of the boundary; and how can the boundary be determined? This section summarizes mapping and boundary identification programs that are used in various states. Second, this section briefly considers the decision in *Phillips Petroleum Co. v. Mississippi,* in which the United States Supreme Court confirmed that under the equal footing doctrine each state joined the Union possessing public trust responsibilities and interests over all lands subject to the ebb and flow of the tide. Third, this section addresses whether the state's jurisdiction under the public trust doctrine reaches beyond the traditional state boundary. Specifically, it looks at examples from New Jersey and California where the courts have, based upon various rationales, extended the state's authority under the public trust doctrine to reach beyond the traditional inland boundary—the mean high-tide line.

The threshold enquiry before considering applications of the public trust doctrine is to determine what lands and waters constitute the "zone" of public trust rights. Almost all coastal states quite explicitly by statute or in case law recognize public trust rights in state waters extending from the former seaward limit of the territorial sea (three nautical miles) to the mean high-tide line, and in navigable rivers, streams, ponds, and lakes to the ordinary high-water mark.[1] A

---

1. More attention has been given in public trust theory to determining the landward extent of the doctrine. With regard to the seaward boundary of the "zone" of public trust rights, a survey of relevant statutes and case law in thirty coastal and Great Lake states indicates the following:

By statute, ten states (Alabama, Alaska, Florida, Illinois, Louisiana, Maine, Massachusetts, Michigan, Mississippi, and Pennsylvania) explicitly apply the public trust doctrine seaward to the three nautical mile limit or to the international boundary. Nine states (California, Delaware, Hawaii, Indiana, North Carolina, Ohio, South Carolina, and Wisconsin) somewhat less certainly define the seaward limit in statutes. Eleven states (Connecticut, Georgia, Maryland, Minnesota, New Hamp-

few states, however, only recognize full public trust protection seaward of the low-tide line.[2]

A state must identify its public trust areas before it can effectively manage them. A clear legal definition of areas subject to the public trust is necessary but does not resolve boundary and title problems, as the Mississippi legislature recognized in its preamble to its public trust tideland mapping law passed in March 1989:

> [a] dispute has developed with respect to lands bordering on tidally affected waters, calling into question title and legal issues believed secured and determined from the date of statehood; that this dispute has cast a doubt and cloud over the title to all the littoral lands and all riparian lands along the rivers and shore lines of the coastal area to such a degree that land sales are being prevented, business and home purchasing has been made difficult or impossible, industrial financing based on such title has become unavailable, and homeowners and other owners have been rendered apprehensive as to their security in their ownership. Economic growth and development in the coastal counties is at a virtual standstill, creating a constantly increasing and incalculable loss of dollars to the area as well as the loss of countless new jobs for the average citizens of our states.[3]

Once it has been established by legislative action or judicial interpretation that the state's trust boundary extends, for example, to the mean high tide, how is the mean high-tide line measured?[4] To accomplish this task, a state may take one of two approaches: it can physically map its public trust lands;[5] it can put the burden on property owners who want to develop their land to make the determination at the permit application stage.

The choice between these two approaches most likely depends upon a state's ability to pay the cost of mapping. Accurately mapping a state's public trust lands

---

shire, New Jersey, New York, Oregon, Rhode Island, Texas, and Washington) apparently do not address the seaward limit issue in statutory law.

In case law, twenty-two states (Alabama, Alaska, California, Connecticut, Florida, Hawaii, Illinois, Indiana, Louisiana, Maryland, Massachusetts, Michigan, Minnesota, Mississippi, New Jersey, New York, North Carolina, Ohio, Oregon, Rhode Island, Texas, and Wisconsin) have applied the doctrine to the three nautical mile limit or to the international boundary in the Great Lakes. Courts in five states (Georgia, Maine, South Carolina, Virginia, and Washington) are somewhat less certain regarding the seaward limit, while in three states (Delaware, New Hampshire, and Pennsylvania) courts have not considered the issue.

Only in Georgia and New Hampshire is there some doubt regarding the seaward limit of the "zone" of public trust rights.

2. These states are Delaware, Maine, Massachusetts, New Hampshire, Pennsylvania, Virginia, and Wisconsin.

3. Miss. Code Ann. 29-15-1 [Supp. 1989. *Ed.*].

4. In Florida, for example, the gradual slope of the lands and the extensive marsh and swamp, estuaries, and freshwater uplands make it extremely difficult to determine the boundaries of public trust lands. See generally *Coastal Petroleum Co. v. American Cyanamid Co.*, 492 So.2d 339 (1986).

5. See, e.g., Miss. Code Ann. 29-15-7-23 (Supp. 1989); Cal. Pub. Res. Code secs. 6331–40 (West Amend. 1989, Supp. 1990).

requires expensive personnel and technical expertise. Once such maps are compiled, however, a state can reliably, uniformly, and efficiently administer a public trust land licensing or leasing program, collecting fees for future projects as well as for existing private uses of public trust land.

In the alternative, a state may put the entire cost of identifying public trust lands onto private property owners, resulting in a pay-as-you-go, incremental mapping process. For example, a state can legislate that all development on public trust land requires a state license. In order to finance any significant development, a private property owner has to satisfy its lender that the property is not subject to such legislation, perhaps by obtaining a statement from the state to this effect. Massachusetts, for example, has established a process under which an applicant may submit a request to the Department of Environmental Protection (DEP) for a "Determination of Applicability," together with a plan of the parcel showing existing fill and structures, references to previous authorizations for existing structures, fill, or dredging, and a delineation of the present high and low watermarks and natural high watermark.[6] After a public notice and comment period, which may include a public hearing, the DEP issues its determination as to whether the parcel is public trust land in recordable form.[7] Under this system, private property owners pay for the determination on individual parcels when necessary to obtain a new license. However, the system does not address existing private uses or smaller new private uses. Another drawback to this system is its ad hoc nature, with determinations dependent upon the quality of a private property owner's historical waterline data.

California provides an example of the advantages and disadvantages of a state's attempt to produce comprehensive public trust lands maps. In 1975 the California legislature charged the State Lands Commission to make an inventory of the location and extent of all ungranted tidelands in the state by the end of 1981.[8] This legislation directed the commission to rely on existing surveys showing the natural tideland boundaries where available and conduct its own surveys where necessary.[9] The commission was to presume the existing ordinary high waterline as the natural line unless "clear and convincing" evidence existed to the contrary.[10] In addition, the commission was to consider whether private owners of long-standing improvements authorized by the government on otherwise public trust lands should receive title to the underlying land.[11]

6. Mass. Regs. Code tit. 310, sec. 9.06 (1990).

7. Id.

8. Cal. Pub. Res. Code sec. 6331.5 (West 1975).

9. Id. Before conducting its own surveys, the commission must submit a report of those ungranted tidelands requiring such surveys along with a cost estimate.

10. Id. at 6332(a) (6) (West Amend. 1981, Supp. 1990).

11. Id. at 6332(a) (4) (West Amend. 1981, Supp. 1990).

Once the commission completed the inventory, it was given eight years to complete a preliminary map of ungranted tideland boundaries.[12] The process worked as follows: the commission would submit a preliminary description of each parcel of ungranted tidelands to the county clerk for the county in which each inventory parcel is located;[13] local agencies and private individuals could protest the preliminary description by alleging incorrect boundary drawings;[14] the commission would either hold a hearing on the dispute or negotiate a boundary agreement that would be binding on the state and all other parties.[15] As its final step, the commission would prepare, adopt, and publish a final Master Map of Ungranted Tideland Boundaries and a Description of Ungranted Tideland Boundaries for each parcel of public trust land in the state.[16] As the regulatory body responsible for leasing state-owned lands, including public trust lands,[17] the State Lands Commission could use the map and descriptions to oversee and influence new development and to require compensation for existing private uses of public trust land.

Unfortunately, the California legislature discontinued funding for this program in 1980, after the commission had submitted its report on the extent of required surveys and costs. California, therefore, currently follows the ad hoc mapping approach. The State Lands Commission's "boundary staff" initially determines whether a particular parcel is public trust land. This determination usually arises as part of either litigation to confirm title to land near the water or the commission's review of applications to lease public trust land for private use.

Mississippi has just begun a similar program but with three important differences. First, Mississippi's mapping statute instructs the secretary of state to depict the public trust land boundary in developed areas as the mean high watermark as of 1973, the effective date of the Mississippi Coastal Wetlands Protection Act.[18] This provision is effectively a conveyance of all public trust lands that were filled prior to 1973, if there was no reservation of public rights.[19] Second, the statute

12. Id. at 6333 (West 1975).
13. Id. at 6334(a) (West 1975).
14. Id. at 6335 (West 1975).
15. Id. at 6336 (West 1975).
16. Id. at 6338 (West 1975). The map and description do not bind upland owners, although boundary agreements do. Id. at 6339(a–b) (West 1975). In New Jersey, on the other hand, almost all activity within mapped coastal wetlands requires a permit. N.J.A.C. 7:7–2.2 (Supp. 1988).
17. Cal. Pub. Res. Code at 6501–9 (West Supp. 1990).
18. Miss. Code Ann. 29–15–7(1) (Supp. 1989).
19. In *Byrd v. Mississippi*, Cause No. 17, 879, Harrison County Chancery Ct. (1989), the Mississippi attorney general brought an action challenging the validity of the conveyance, arguing that the Mississippi Constitution prohibits gifts of property absent specific legislative findings. The Mississippi Supreme Court recently upheld the constitutionality of the statute. *Byrd v. Mississippi*, Ms. Chanc. Ct., Harrison Cty., 2d Dist., No. 17, 879 (April 18, 1990).

expressly directs the secretary of state to notify property owners whose "lands are subject to the public trust and are in violation of such trust."[20] If no resolution is reached within six months of the receipt of such notice, the state attorney general has authority to bring a title confirmation suit to resolve ownership of the property.[21] The state has the burden of proving the land is subject to the public trust.[22] Finally, the Mississippi mapping program is funded directly from the revenues generated by the state's public trust land leasing programs.[23]

1. The Phillips Petroleum Decision and the Geographic Scope of the Public Trust Doctrine

In 1988 the Supreme Court of the United States clarified the geographic scope of the public trust doctrine in the *Phillips Petroleum* case.[24] The Court restated the equal footing doctrine under which each new state entered the Union with the same rights as the original states possessed in public trust areas. The Court examined nineteenth-century precedents for guidance in applying the public trust doctrine in the United States. These decisions interpreted the public trust doctrine under English common law to cover lands beneath tidal waters, whether navigable or not, and held that title to such areas passed to the new states under the equal footing doctrine.[25] Later the Supreme Court recognized that the public trust doctrine in the United States, adapting to the different physical conditions in the United States as compared to England, which has no great inland rivers and water bodies, and to preserve important public interests in navigation and commerce, included navigable freshwater bodies.[26] Based upon these long-standing precedents, the Court affirmed that the states "owned all the soil beneath waters affected by the tides."[27]

The Court acknowledged that the scope and content of the public trust doctrine has varied from state to state, explaining that although all states began with the same rights in public trust areas, they may "define the limits of the lands held in public trust and . . . recognize private rights in such lands as they see fit."[28]

20. Miss. Code Ann. 29–15–7(4) (Supp. 1989).
21. Miss. Code Ann. 29–15–7(5) (Supp. 1989).
22. Id.
23. Miss. Code Ann. 29–15–9 (Supp. 1989).
24. *Phillips Petroleum Co. v. Mississippi*, 484 U.S. 469 (1988).
25. Id. at 477–78. Phillips Petroleum and petitioners argued that, because practically all navigable rivers in England are tidally influenced, English law used the terms "tidewater" and "navigability" synonymously; but in the petitioners' view, "navigability" rather than tidal influence was the critical factor in applying the public trust doctrine. Therefore, states acquired no interest in non-navigable tidal waters under the doctrine.
26. See, e.g., *Barney v. Keokuk*, 94 U.S. 324 (1877).
27. *Phillips Petroleum*, 484 U.S. at 479.
28. Id. at 475 [citing *Shively v. Bowlby*, 152 U.S. 1, 26 (1894)].

Thus states may narrow or expand the area of public trust lands and waters. In several states, by common law or by statute, a littoral property owner holds the fee extending to the mean low waterline. But such property owners hold their fee subject to the public's rights to use the property for trust purposes. In some states, property owners may hold an absolute fee. The basic rule recognized in *Phillips Petroleum,* however, remains the same: a state may expand or narrow the "zone" of public trust interests if it follows appropriate procedures to protect public trust interests.[29]

2. Expanding the Doctrine beyond Its Traditional Boundaries

In recent years, the geographic scope of the public trust doctrine has been "expanded" in some states by extending its authority to lands and resources above the mean high tidemark or ordinary high watermark. The boundary of public trust areas per se has not been extended in such states; rather, courts have recognized that the state, as trustee, has the right (or the obligation) to protect its public trust lands, waters, and resources by regulating activities outside public trust areas that may affect them and by requiring access across privately owned uplands to public trust lands.

For example, in New Jersey, the state supreme court has held that the public must have reasonable access to the dry sand areas of private beaches for passage and recreational use as well as to the dry sand areas of municipal beaches.[30] The New Jersey courts have determined that the reasonable enjoyment of the public's

29. See *Phillips Petroleum* 484 U.S. at 476, 482; *Mono Lake,* 33 Cal.3d 419, 189 Cal. Rptr. 346, 658 P.2d 709. For example, a state may decide (by statute, judicial decision, or state agency determination) that certain trust lands due to historical filling, their land-locked condition, or location are no longer appropriate for trust purposes and their regulation under the public trust is not practicable or desirable. See, e.g., *City of Berkeley v. Superior Court of Alameda County,* 162 Cal. Rptr. 327, 606 P.2d 362 (1980).

30. E.g., *Matthews v. Bay Head Improvement Ass'n, Inc.;* 95 N.J. 306, 471 A.2d 355 (1984). *Bay Head Improvement* extended the New Jersey court's earlier decision in *Van Ness v. Borough of Deal,* 78 N.J. 174, 393 A.2d 571 (1978), which held that the public trust doctrine required municipally owned dry sand areas adjacent to tidal waters open to both residents and nonresidents on an equal basis (although nonresidents would be required to pay the same fee as paid by residents). In a series of user fee cases, the same court implied that this right of access to the beach included a corollary meaningful opportunity to enjoy the beach, free of restrictions intended to discourage use. *Bay Head Improvement* involved a suit by the New Jersey Public Advocate along with private individuals against a quasi-public body, the Bay Head Improvement Association, which was restricting access to a beach held in trust by the association. The court held that the association's restrictive policies frustrated the public's right to access the tidelands. Significantly, in dicta, the court made the logical extension between recognition of a public trust beach and the right of access to the beach even if that meant crossing private property. "Without some means of access," the court wrote, "the public right to use the fore shore would be meaningless. . . ." This does not mean the public has an unrestricted right to cross at will over any and all property bordering on the common property. The public interest is satisfied so long as there is reasonable access to the sea. Id.

right to use tidelands, guaranteed by the public trust doctrine, includes the enjoyment of the dry sand areas. In so holding, the New Jersey courts have identified a number of factors to be considered in extending access to trust lands. These factors include the extent of existing access to the water, the extent of publicly available dry sand areas nearby, and the degree of public demand for beach use.

The California Supreme Court has held that the state, as trustee of its public trust lands, may regulate the use of non-navigable waterways when such use affects public trust waters such as Mono Lake.[31] The court held that the state could regulate the diversion of waters from tributary streams to ensure that such diversion would not adversely affect the public trust waters into which the streams flowed. The rationale used by the California court is similar to that of the New Jersey court: the public trust doctrine can be applied by the governmental trustee to reach beyond the doctrine's traditional boundaries to the extent necessary to ensure that trust lands, waters, and resources are protected or that the public may have reasonable access to and use of them. In *Mono Lake*, the court's reasoning also echoes the common law doctrine of nuisance.[32] Because the public trust doctrine is an ownership-based doctrine, the state as trustee may prevent upland owners from committing a nuisance or conducting a "nuisance-like" activity (i.e., an activity not classifiable as a common law nuisance but which has similar adverse effects) that will injure trust interests. In *Mono Lake*, the "nuisance

31. *Nat'l Audubon Soc'y v. Superior Court of Alpine County,* 33 Cal.3d 419, 189 Cal. Rptr. 346, 658 P.2d 709, *cert. den'd,* 464 U.S. 977 (1983) (called *Mono Lake*). The *Mono Lake* decision was based upon early established principles giving the state authority to protect its resources from destruction or pollution by activities upland or upstream. See, e.g., *People v. Gold Run Ditch & Mining Co.,* 66 Cal. 138, 146–47, 151–52, 4 P. 1152 (1884); *People v. Russ,* 132 Cal. 102, 106, 64 P. 111 (1901).

32. An owner of land may recover for private nuisance against another for nontrespassory invasion of the owner's land, as long as the "invasion" is deemed unreasonable. *Restatement (Second) of Torts* 831D, 831E. Pollution of waters or interference with the flow of waters may constitute a nuisance. Id. at 832, 833. The law of public nuisance may also be applicable. By statute or judicial decision in each state, a person is liable for any act or omission which unreasonably obstructs, interferes with, or damages the rights, interests, comfort, or convenience of the community at large or the general public, including conducting activities on one's land leading to noise, obnoxious odors, smoke, dust, or pollution affecting the public (e.g., a pigpen or blasting operations) or interfering with the exercise of a public right (e.g., obstructing a public highway). A public official, usually the attorney general, is empowered to "abate" such nuisances, typically by instituting a court action seeking an order to enter upon the land from which the nuisance is emanating to terminate it. See generally, Prosser and Keeton, *The Law of Torts,* ch. 17 (5th ed. 1984). The law of nuisance is an extension of the ancient common law principle *sic utere tuo ut alienum non laedas* ("use what is yours so that others are not injured"), which for centuries has been a basic maxim underlying the regulation of landowners who share the same physical environment and which demonstrates that the use of private land has never been, in our law, an entirely private matter beyond regulatory authority. See e.g., *Mugler v. Kansas,* 123 U.S. 623, 665 (1887); *Penn Central Transportation Co. v. City of New York,* 438 U.S. 104, 125 (1978); and ch. 2, sec. H below.

equivalent" was the diversion of water from the lake's tributaries, which would have lowered the lake's level and endangered its plant and wildlife.

Regulating or restricting upland activity to protect and preserve public trust areas based on public trust principles and the state's obligations as trustee to manage such areas should prove particularly useful in protecting and preserving coastal bays and harbors in densely populated areas, such as the Chesapeake Bay, Boston Harbor, and San Francisco Bay. Such areas are subject to pollution from upland activities that threaten their survival. The public trust doctrine, employed together with the state's traditional police power authority, is a strong weapon in the arsenal to defend these sensitive and critical lands, waters, and resources.[33]

B. PUBLIC TRUST USES

After determining the boundaries of public trust areas, what uses are permitted, restricted, or protected in them? This section first provides an overview of the current law regarding appropriate public trust uses in various states and then considers whether some uses are "protected" under the public trust doctrine in order to preserve public trust lands, waters, and resources from harmful or consumptive activities. Next, the section examines one of the more difficult tasks facing states and coastal management agencies: how to choose among appropriate but competing uses for public trust areas and resources. This section recommends that state legislatures and their agency delegates should take a more aggressive role in defining and choosing among appropriate public trust uses of public trust areas and describes approaches that a state agency might take to establish a framework for analyzing and selecting among competing trust uses.

Recurring problems in the application of the public trust doctrine have been defining appropriate or protected uses for public trust lands, waters, and resources and deciding among competing appropriate or protected uses. As described below, these problems typically have been addressed by the courts on an ad hoc basis, resulting in considerable variation among the states. In response, state legislatures and their agency delegates may designate specific appropriate and protected uses of public trust lands, waters, and resources. While such an active role will not eliminate the problem of choosing among competing uses, it will place the problem in the hands of an elected body (or its delegate) more accustomed than the courts to balancing the public's interests and more sensitive and experienced than the courts in accommodating the legitimate demands of competing groups.

33. For a discussion of comprehensive planning using the public trust doctrine, see ch. 3, secs. A and B. For a discussion of marinas and harbors and the role of the public trust doctrine, *see* ch. 3, sec. C, subsection 7.

The risk of this approach is that state legislatures and their delegates may be less responsive than courts to the interests of future beneficiaries—i.e., future generations—because future generations do not vote in present-day elections. This risk may be minimized by either the creation of private rights of action on behalf of beneficiaries[34] or an expansion of the doctrine expressed in *City of Berkeley v. Superior Court of Alameda County*,[35] declaring that a state legislature may not exercise its powers to give away the discretion of its successor legislatures such that future legislatures have no trust resources over which to exercise their discretion.

1. Current Law: Appropriate Public Trust Uses

The public trust doctrine has traditionally protected the public's rights to fishing, commerce, and navigation, rights stemming from both Roman and English law, which protected the public's use of coastal lands, waters, and resources essential to improving its livelihood as well as to assuring survival. However, since the adoption of the public trust doctrine in the United States, many state courts have responded to changing public needs and demands by expanding the list of uses that are protected by the public trust doctrine.

Courts in California and New Jersey have been among the most active in deciding public trust disputes and have therefore had the greatest opportunity to consider nontraditional uses and to expand the list of uses that are either pro-tected or appropriate on public trust lands. In California the expansion of uses beyond those traditionally recognized was first explicitly set out in *Marks v. Whitney*,[36] in which the court observed that the public trust uses were not limited to navigation, commerce, and fisheries: "Public trust easements are traditionally defined in terms of navigation, commerce and fisheries. They have been held to include the right to fish, hunt, bathe, swim, to use for boating and general recreation purposes, the navigable waters of the state, and to use the bottom of the navigable waters for anchoring, standing or other purposes. The public has the same rights in and to tidelands."[37]

The court also recognized preservation as an appropriate "use":

The public uses to which tidelands are subject are sufficiently flexible to encompass changing public needs. In administering the trust the state is not burdened with an outmoded classification favoring one mode of utilization over another. There is a growing public recognition that one of the most important public uses of tidelands—a

34. See ch. 2, sec. C.

35. 26 Cal.3d 515, 162 Cal. Rptr. 327, 606 P.2d 362 (1980), *cert. den'd*, 449 U.S. 840 (1980).

36. 6 Cal.3d 251, 98 Cal. Rptr. 790, 491 P.2d 374 (1971). As described in chapter 1, the public trust doctrine as adopted in the United States is a common law doctrine that has been developed almost entirely through court decisions rather than through acts of state legislatures.

37. *Marks*, 6 Cal.3d at 259, 98 Cal. Rptr. at 796, 491 P.2d at 380 (citations omitted).

use encompassed within the tidelands trust—is the preservation of those lands in their natural state, so that they may serve as ecological units for scientific study, as open space, and as environments which provide food and habitat for birds and marine life, and which favorably affect the scenery and climate of the area.[38]

California courts have followed the rationale of *Marks v. Whitney* to extend the public trust doctrine in other circumstances.[39]

Similarly, the courts in New Jersey have also recognized that valid public trust uses include bathing, recreation, swimming, and other shore activities. For example, in *Borough of Neptune City v. Borough of Avon-by-the-Sea,*[40] the court noted:

> We have no difficulty in finding that, in this latter half of the twentieth century, the public rights in tidal lands are not limited to the ancient prerogatives of navigation and fishing, but extend as well to recreational uses, including bathing, swimming and other shore activities. The public trust doctrine, like all common law principles, should not be considered fixed or static, but should be molded and extended to meet the changing conditions and needs of the public it was created to benefit.[41]

In Wisconsin the court has extended the public trust uses to include waterskiing.[42] The court first analyzed earlier cases and determined that, in Wisconsin, the scope of appropriate and protected public trust uses had been expanded to include recreation. It then stated that waterskiing was encompassed within the scope of recreation. In a more recent case, the Idaho Supreme Court echoed the trend described above, stating that "[i]t is now held that the [public] trust protects varied recreational uses in navigable waters . . . [t]he trust is a dynamic, rather than static, concept and seems destined to expand with the development and recognition of new public uses."[43]

By contrast, the Maine Supreme Judicial Court held that public trust uses in privately owned trust lands are limited to fishing, fowling, and navigation and do not include bathing, swimming, and other shore activities.[44] In Massachusetts, courts have held that while the public trust doctrine includes the rights to seek or take fish, shellfish, fowl, and marine plants from a vessel or on foot and the right to pass freely over privately held tidelands for the purposes of fishing and fowling

38. Id. at 259–60, 98 Cal. Rptr. at 796, 491 P.2d at 380 (citation omitted).

39. *Mono Lake*, 33 Cal.3d 419, 189 Cal. Rptr. 346, 658 P.2d 709; *Carstens v. California Coastal Comm'n*, 182 Cal. App. 3d 277, 227 Cal. Rptr. 135 (1986).

40. 61 N.J. 296, 294 A.2d 47 (1972).

41. Id. at 390, 294 A.2d at 54. See also *Matthews v. Bay Head Improvement Ass'n*, 95 N.J. 306, 471 A.2d 355 (1984); *Van Ness v. Borough of Deal*, 78 N.J. 174, 393 A.2d 571 (1978).

42. *State v. Village of Lake Delton*, 93 Wis.2d 78, 286 N.W.2d 622 (1979).

43. *Kootenai Environmental Alliance, Inc. v. Panhandle Yacht Club, Inc.*, 105 Idaho 622, 671 P.2d 1085, 1088 (1983).

44. *Bell v. Town of Wells*, 557 A.2d 168 (Me. 1989) (called *Wells Beach*).

(the "natural" derivatives of the historical rights of navigation), the public's rights do not include the right to pass over private property for any purpose other than fishing or fowling, the right to bathe or swim on private beaches, or the right to take soil from public areas.[45]

In addition to efforts to expand the number of appropriate and protected public trust uses beyond the traditional uses (fishing, fowling, and navigation), state legislatures and their agency delegates may also consider the range of contemporary uses and activities that have developed from or in additional to these traditional uses.[46]

2. Protected Uses

In addition to the question of what uses are permitted on public trusts lands, state courts have also considered whether the public trust doctrine protects trust lands, waters, and resources against certain uses. For example, in addition to expanding public trust uses to include recreation, boating, and swimming, California has applied the public trust doctrine as a tool to protect the environment affirmatively.

In *Marks,* described above, for example, the court recognized that public trust uses included preservation of public trust lands "in their natural state." More recently, in *Mono Lake*[47] the same court was asked to review water licenses under the rubric of the public trust doctrine. The water licenses had been issued under the state's prior appropriation doctrine to divert water to the city of Los Angeles from contributory streams leading to Mono Lake, a large navigable lake. The

45. See, e.g., *Opinion of the Justices,* 365 Mass. 681, 313 N.E.2d 561 (1974); *Butler v. Attorney General,* 195 Mass. 79, 80 N.E. 688 (1907).

46. An important question, not yet clearly addressed by the courts, is how far a coastal management agency may go, for example, in its leasing or licensing programs, in permitting uses that are not a part of or related to the more traditional trust uses, when state statutes and court decisions are silent as to the scope of trust uses. Given the almost universal acceptance of the doctrine of implied agency powers (see ch. 2, sec. D) and the widely accepted concept of "accessory uses" in the zoning field— where it is generally understood to refer to any "use which is clearly incidental to and customarily found in connection with and located on the same zoning lot as is the principal use to which it is related" [Rathkopf, *The Law of Zoning and Planning,* 23–2 (4th ed. 1986)]—it would appear that an agency could go beyond traditional uses, at least when the permitted activities are both accessory and water dependent (e.g., allowing the construction of a food and marine supply store on or near a wharf so as to facilitate boating and fishing). Indeed, the agency could probably permit even a wholly private, non-water-dependent use when to do so would produce significant public benefit, so as to increase the public's use of the waterfront (e.g., allowing a private developer to construct a marina and condominium in any area, such as decaying warehouses on a rotting pier, that would otherwise provide no public access and in fact create a condition endangering public safety, at least in circumstances where there are no public funds available to repair the pier and the developer agrees to conditions that would enhance public trust rights such as boating and fishing access).

47. 33 Cal.3d 419, 189 Cal. Rptr. 346, 658 P.2d 709 (1983).

evidence indicated that the loss of water led to increased salinity of the lake, posing a risk to its brine shrimp, birds, and other wildlife. The court found that the public trust doctrine not only limited the uses to which Mono Lake could be put, but also held that the state's duties as trustee included an obligation to preserve the trust property. Specifically, the court held that "the public trust is more than an affirmation of a state power to use public property for public purposes. It is an affirmation of the duty of the state to protect the people's common heritage of streams, lakes, marshland and tidelands, surrendering that right of protection only in rare cases when the abandonment of that right is consistent with the purposes of the trust."[48]

This case is particularly significant because it pitted the long-standing prior appropriation doctrine in water rights law against the public trust doctrine and held that the state had authority and responsibility to reconsider such prior rights pursuant to the public trust doctrine. The court refrained from ruling that either doctrine took precedence over the other, but clearly held that, despite the prior appropriation doctrine, allocating water must take full and adequate account of the public trust interests at stake.[49]

The shift in thinking from permitting certain uses on public trust lands to affirmatively protecting natural resources and uses on public trust property is a significant change in the public trust doctrine's traditional focus. Originally, the doctrine permitted members of the public to use public trust resources for the purposes of navigation and to have access for fishing and fowling. The doctrine opened trust lands for exploitation or use by all members of the public—not just by those who owned the land. Fishing, commerce, and navigation were "protected" because they provided critical necessities of life at the time of the doctrine's inception. There is little evidence that the original purpose of the doctrine was to preserve trust resources, for example, by assuring that there would always be fish in the waters for fishing or that the waters would be kept unpolluted.[50]

---

48. Id. at 441, 189 Cal. Rptr. at 360–61, 658 P.2d at 724.

49. See also *United Plainsmen Ass'n v. North Dakota State Water Conservation Comm'n*, 247 N.W.2d 457 (N.D. 1976), in which the North Dakota court held that the discretionary authority of state officials to allocate water resources is explicitly circumscribed by the public trust doctrine, which requires, at a minimum, a determination of the potential effect of allocation of present water supply and future water needs on public trust interests.

50. See Huffman, "Trusting the Public Trust Interest to Judges: A Comment on the Public Trust Writings of Professors Sax, Wilkinson, Dunning and Johnson," 63 Denver U. L. Rev. 565 (1986). But see Note, "The Public Trust in Tidal Areas: A Sometimes Submerged Traditional Doctrine," 79 Yale L. J. 762, 785–86 (1969–70), which argues to the contrary: "[a]lmost all the elements of a public trust right of conservation in tidal areas have long existed, because it has long been recognized that adequate conservation is a necessary prerequisite to the enjoyment of protected activities." Such lack of evidence undoubtedly reflects the general attitude, until recently, that nature's bounty was inexhaustible rather than a considered restriction of the public trust doctrine. Had the issue been squarely put before those concerned with the public trust, it is virtually certain that they would have

Thus, to the extent courts have integrated the doctrine by requiring states as trustees to protect the natural resources held subject to the public trust, state agencies and coastal managers will be in a position to use the public trust doctrine as a tool in coastal resource protection and management.

3. Competing Uses

A continuing problem in the application of the public trust doctrine concerns the choice among competing public trust uses. How should the courts, the legislature as trustee, or the public trust agency (if the legislature and the courts have not spoken) choose among multiple uses, all of which are appropriate uses under the state's public trust doctrine, but some of which may be incompatible? For example, how should the courts or the legislature choose between potential development of public trust land as a public marina, preservation of the public trust land in its natural state for future generations, use of the public trust lands to explore for and extract mineral resources, and use of public trust lands for the development of fisheries? Is there some "hierarchy" of uses that state courts, legislatures, or administrators can systematically apply in their decision-making process to determine which public trust use should prevail? The courts have offered some guidance but have stopped short of announcing a specific hierarchy of uses or similar formula. Some state agencies have stepped in where the courts have not and have adopted rules and regulations establishing a hierarchy of uses for specific areas.[51]

While a number of state courts have held that the public trust doctrine does not prevent a state from selecting between appropriate or protected trust uses, their decisions have not been very clear as to the appropriate mechanisms for choosing among those uses. For example, in *Mono Lake,* the California court considered the clash between valid public trust uses and incompatible rights under the prior appropriations doctrine. The court explicitly stated that the public trust doctrine does not prevent a state from choosing among trust uses.[52] Furthermore, the court rejected the attorney general's encompassing argument that all public uses are valid trust uses that must be protected. Instead, the court stated that trust uses must relate directly to uses and activities in the vicinity of the lake, stream, or tidal reach at issue, and that they must be balanced when considering other uses which may interfere with public trust uses.

Another California court addressed the argument that the state's coastal com-

---

seen it as an inherent part of the public trust to take steps to preserve the public's access to the coastal resources, because the evils thereby created were precisely the same as would have existed had private landowners monopolized the shores and excluded the public. See generally, ch. 3, sec. C, subsection 5.

51. See ch. 3, secs. A and B.

52. *Mono Lake,* 33 Cal.3d at 440, 189 Cal. Rptr. at 360, 658 P.2d at 723 (citations omitted).

mission, as trustee, was responsible to preserve recreational access to tidelands without consideration of other competing public trust purposes.[53] The court again rejected this argument holding that "the public trust doctrine, as codified in the California Constitution does not prevent the state from preferring one trust use from another."[54] The court cited a number of earlier cases in which California courts had permitted the issuance of permits for such uses as oil and gas exploration, restaurants and retail businesses, and a freeway bridge. The court's decision, which was grounded in the California Constitution and the California Coastal Act, stated:

> We find nothing in Article X, Section 4 to preclude the Commission from considering commerce as well as recreational and environmental needs in carrying out the public trust doctrine.
>
> The provisions of the California Coastal Act of 1976 demonstrates the Legislature's recognition of the need to balance competing interests in the need to preserve the coastal zone. Maximization of public access is one of several goals articulated in the Act.[55]

Similarly, in Oregon the Supreme Court balanced the community need for an airport expansion against the potential interference with public uses for water-related purposes in connection with a proposal to fill an estuary to expand an airport runway.[56] The court held that the public trust doctrine does not prevent all filling for non-water-related uses. However, it found that in the specific case, the director of the Division of State Lands had not taken the appropriate steps to balance the need for the airport extension against the valid public trust interests in the site, and therefore the court struck down the permit.

The clear theme to be drawn from these judicial decisions is that, while the courts have not established any specific hierarchy of uses, they have made it clear that state agencies must balance public trust interests against other interests before permitting any use of public trust resources. Given this general mandate, state agencies with the authority to act as trustees[57] may develop their own hierarchy of appropriate uses for public trust lands and resources. If the agency develops this hierarchy in accordance with the general principles of administrative law and rulemaking, so that its choices are the product of procedurally sound rulemaking or adjudication and clearly articulate an analysis which weighs public benefits against public detriments,[58] its plan will be given deference by the courts and will in all likelihood be safe from successful legal challenge.

53. *Carstens,* 182 Cal. App. 3d 277, 227 Cal. Rptr. 135 (1986).

54. Id. at 289, 227 Cal. Rptr. at 143 (citing *Mono Lake,* 33 Cal.3d at 439 n.21, 189 Cal. Rptr. at 359 n.21, 658 P.2d at 723–24 n.21).

55. Id. at 289–90, 227 Cal. Rptr. at 143.

56. *Morse v. Oregon Division of State Lands,* 285 Or. 197, 590 P.2d 709 (1978).

57. See ch. 2, sec. D.

58. See ch. 2, sec. D.

One consistent factor in developing a hierarchy of uses is water dependency. It seems clear, based upon state programs and upon state court decisions, that uses of public trust lands and resources that stress water dependency (i.e., uses that in fact require access to the shore) will be favored over non-water-dependent uses, even if both are recognized as appropriate uses under the state's public trust doctrine.[59] However, as the above cases suggest, a state's choice among competing public trust uses is not governed by a mechanical formula. Instead, courts are likely to uphold a state's choice of use so long as there is evidence that the state or its agency delegate has given careful consideration to competing uses and has taken steps to balance those competing uses against the use that is eventually selected.[60]

4. Expanding Role of State Legislatures and Their Delegates in Defining and Choosing among Public Trust Uses

The analysis in *Mono Lake*, together with the general trend followed by courts in requiring that legislatures and state agencies specifically address public trust concerns in preferring one use over another, suggests the potential for a more active role on the part of legislatures and administrative agencies in defining and choosing among public trust uses. This requirement raises the question whether a legislature should define a prioritized list of public trust uses to be applied in each instance of conflict or whether each instance should be considered by the public trust agency on an ad hoc basis, weighing all of the interests and uses involved. In the absence of such a list, the courts have provided state legislatures and their agency delegates with some guidance in choosing among uses. Both the case law and state statutes establish a priority for uses closely resembling the traditional trust uses—fishing, commerce, and navigation—and also for water-dependent uses over non-water-dependent uses.

As discussed in chapter 2, section D, if a state agency follows the general principles of administrative law and has authority to act as trustee under the public trust doctrine, then the courts will normally defer to the agency's rule-making in any legal challenge. Thus a state legislature or its agency delegate could

59. This widely accepted preference for water-dependent uses arises out of the common theme among the traditional public trust uses (fishing, fowling, and navigation) which are water dependent and has been codified in the Massachusetts Chapter 91 program (see ch. 5). Water dependency has been used as the linchpin for other states' coastal management programs, e.g., North Carolina's Coastal Area Management Act of 1974, 113A–100 et seq.

60. The question of valid public trust uses and the hierarchy of uses also becomes critical in determining whether or not the state has effectively conveyed public trust lands free from public trust interest. Specifically, a number of state courts have held that where property is conveyed to a private party, and where that conveyance has not been made for purposes which are valid under the public trust doctrine, such conveyance has been made (and the property remains) subject to the public trust. See discussion, ch. 2, sec. F.

take any of three approaches in establishing a framework for analyzing and choosing among competing trust uses:

1. establish a list of specific uses, identifying uses that will be afforded priority in certain public trust lands;
2. decide each case individually, using general public trust principles relevant to the coast; or[61]
3. establish a set of criteria to govern the process of choosing among competing uses. The criteria might include, for example:
    a. favoring water-dependent uses over non-water-dependent uses,
    b. favoring uses that have been recognized as appropriate under the public trust doctrine in the state over other uses,
    c. favoring water-related or water-oriented uses over non-water-related and non-water-oriented uses,
    d. favoring those uses that most effectively enhance public access to the shore, and
    e. favoring those uses that provide the greatest public benefit, in terms of numbers of people affected or societal values served.[62]

C. ANALOGIES TO PRIVATE AND CHARITABLE TRUST LAW

This section recognizes that there is limited statutory or case law prescribing the rights and obligations of the state or state agency which acts as trustee over public trust lands and resources. This section recommends that states examine the law of private and charitable trusts as an analogous body of law which can provide guidance in defining trusteeship responsibilities. Private and charitable trust law is, with little difficulty, adaptable to public trust issues. Thus this section discusses the various rights and obligations of private trustees and how such rights and obligations can inform decisions by state trustees exercising their authority under the public trust doctrine.

Having reviewed the state's law regarding the boundaries of public trust land and the uses permitted or protected on trust lands, the next question is, what does it mean to hold this land "in trust" or to be a "trustee"? Despite its fundamental character, this question has not been addressed squarely by the courts. Although the United States Supreme Court has declared that states hold submerged and navigable lands subject to the trust, few courts have taken the next

---

61. For example, water-dependent uses would have priority over non-water-dependent uses.

62. The need to establish criteria stressing the priority of water-dependent uses is probably most urgent in connection with the management of waterfront development in urban harbor areas. For a further discussion of this issue, see ch. 3, sec. C, subsection 7.

step and identified the actual "trustee" or defined with any specificity the trust-ee's duties and responsibilities.

Because few courts have answered these questions in the public trust context, this section recommends that state lawmakers and agencies should look to private and charitable trust law as a potentially useful source of analogous law for assistance in addressing issues of public trust implementation for which legislative and judicial precedents are lacking.[63] The vast body of law and practical experience under general private and charitable trust law has yet to be tapped in formulating public trust principles. Specifically, if a state is a trustee over public trust lands and resources for the benefit of its people,[64] then the imposition of the general rights and obligations of a private or charitable trustee establishes useful parameters and guidelines for state courts and agencies to use to evaluate the appropriate role of the state in protecting and using its trust property.[65]

The private trust analogy also focuses on a number of serious policy questions, particularly the evident conflict between the interests and uses of the present-day beneficiaries and the interests and uses of future generations. While the analogy to private trust law does not necessarily provide a specific resolution of these issues, it can provide courts and agencies with guidance that may be applied in making difficult choices between current and future beneficiaries, as well as among current beneficiaries.

63. Under the law of virtually every state, a trust is a fiduciary relationship with respect to property (which clearly would apply to property-based public trust interests), by which a trustee is empowered to deal with the property for the benefit of another person or persons (the beneficiaries). A private trust typically arises as a result of the direction of one person (the settlor) to create it. A charitable trust is a particular type of trust designed for the benefit of a class of people or for the public generally, which is created for charitable, educational, religious, or scientific purposes and which is enforceable on behalf of the public by the state's attorney general.

64. Because a single person or entity cannot be both sole trustee and sole beneficiary of a trust [Restatement (Second) of Trusts 115(5) (1959)], it might at first glance be thought that trust principles are inapplicable, in the American system, to public trust analysis if the people are in effect both the trustee and the beneficiary. This is, however, a simplistic and inaccurate characterization of our democratic system because "the people" is not a unitary concept. It does not require all of the people (or their representatives) to enact laws or pass constitutional amendments but only a simple or other majority of the people. Consequently, the proper analogy is that of several trustees who may also be all of the beneficiaries of the trust [Id., 115(4)]. Thus no conceptual difficulty arises in applying trust principles in the public trust context. Yet another perfectly acceptable analogy is that the sovereign people have created the public trust for themselves as beneficiaries and their agent, the legislature, as the trustee. Id. 114, 127.

65. In fact, one of the few cases to consider the responsibilities of the trustee is Slocum v. Borough of Belmar, 238 N.J. Super 179, 569 A.2d 312 (1989), in which the court held that the Borough of Belmar was "a trustee over its beach area" and that "[a] public trustee is endowed with the same duties and obligations as an ordinary trustee." 569 A.2d at 317. The court then drew on private trust law to hold that Belmar had violated its duty of loyalty to the public by increasing beach admission fees. Id.

Another fruitful area of law and practical experience may be found by analogizing public trust obligations and responsibilities to the charitable trust. Charitable trusts are generally subject to the law governing private trusts. In addition, however, they have been adapted to meet the special requirements of charitable trusts, many of which are analogous to the public trust. A particular benefit of the use of this analogy is the potential for the imposition of the *cy pres* doctrine to public trust lands.[66]

1. Judicial Statements Declaring the State as Trustee

There are numerous examples of courts declaring that the state is the trustee or holds lands in trust, but none adequately defines that role. In the watershed public trust case of *Illinois Central*,[67] for example, the United States Supreme Court considered two questions. First, is the public trust doctrine applicable to inland waters? And second, could the state's public trust interests be terminated by the grant of public trust lands to a private party? The Court held that the state as trustee could not abdicate its trusteeship responsibilities over public trust lands.[68] However, the Court did not define the rights and obligations of the state as trustee, other than the right to revoke its improvident conveyance of the public trust land.

Similarly, in *City of Berkeley*,[69] the Supreme Court of California made the following statement: "Even before *Illinois Central* was decided it was recognized in California that the state had the authority as administrator of the trust on behalf of the public to dispose absolutely of title to tidelands to private persons if the purpose of the conveyance was to promote navigation and commerce."[70] The court then considered the issue of conveyance of public trust lands. In 1870 the California Board of Tideland Commissioners conveyed title to tidelands under San Francisco Bay to private purchasers, granting them all right, title, and interest of the state of California in and to the property. Overturning earlier decisions affirming the validity of conveyances under the 1870 statute, the court held that submerged and tidal lands conveyed by board deeds remain subject to the public trust.[71] The court did not, however, address the trustee's duties with respect to trust lands where conveyancing is not the specific issue.[72]

66. See ch. 2, sec. C, subsection 5.
67. 146 U.S. 387 (1892).
68. Id. at 452–53.
69. 26 Cal.3d 515, 162 Cal. Rptr. 327, 606 P.2d 362.
70. Id. at 523, 162 Cal. Rptr. at 331, 606 P.2d at 366.
71. Id. at 534–35, 162 Cal. Rptr. at 338, 606 P.2d at 373. The court examined the 1870 statute in light of the criteria it had set forth in *People v. California Fish Co.*, 166 Cal. 576, 138 P. 79: statutes purporting to abandon the public trust are to be strictly construed; the intent to abandon must be

The Supreme Judicial Court of Maine was presented with a unique opportunity to review these issues in 1981.[73] The court was called upon to review a proposed statute which would declare that title to all tidal lands in Maine filled before 1975 be deemed to be owned free of the public trust. The first two questions posed to the court by the legislature were straightforward trust questions, which, unfortunately, the court did not answer in its decision:

(1) does the state of Maine have a trust responsibility for the benefit of the people of Maine in lands which are now or were formerly submerged under the territorial waters and great ponds, or in lands which are now or were formerly intertidal lands? and

(2) if the answer to question (1) is in the affirmative, what are the rights of the beneficiaries and responsibilities of the trustee with respect to the filled, submerged and intertidal lands impressed with the trust?[74]

Rather than respond to these general trust questions, the court focused on the third and fourth questions:

(3) would [the statute], if enacted into law, be invalid as exceeding the authority of the Legislature under the . . . Maine Constitution; and

(4) would the statute, if enacted into law, be invalid as violative of the state's legal responsibilities as trustee for the public of the submerged and intertidal lands located in tidal waters and great ponds on Maine?[75]

The court found that the lands were "impressed" with the public trust, but decided that the state legislature had taken the necessary steps to convey its interest lawfully.

Thus the courts have not clearly addressed two critical questions: what entity within the state is the trustee, and what are the responsibilities and obligations of the state as trustee? This section recommends responses to these questions, looking first to the origins of the public trust doctrine and then analogizing to well-established trust principles in private and charitable trust law.

---

clearly expressed; any reasonable interpretation of the statute that will preserve the trust should be followed (Id. at 369).

72. *Marks,* 6 Cal.3d 251, 260, 98 Cal. Rptr. 790, 796, 491 P.2d 374, 380 (1971) (citations omitted); see also *Thomas v. Sanders,* 65 Ohio App. 2d 5, 413 N.E.2d 1224 (1979); *Save Ourselves, Inc. v. Environmental Control Comm'n,* 20 ERC 214 (La. 1984) (state agency reviewing hazardous waste site assignment on submerged lands is duty bound, as trustee over submerged lands, to consider alternate projects, alternate sites, and mitigation measures and to quantify environmental costs of facility siting on submerged lands against social and economic benefits of project).

73. *Opinion of the Justices,* 437 A.2d 597 (Me. 1981). In Maine, Massachusetts, and a few other states, the state constitution authorizes the state supreme court to render advisory opinions to the legislature or the governor on the constitutionality or validity of pending legislation.

74. Id. at 600.

75. Id.

2. Identification of the Trustee

In private trust law, the trustee is named in the trust instrument. The source of confusion in identifying the trustee under the public trust doctrine stems from the English common law distinction between the Crown and the people. When the United States was formed, the states adopted the English common law. However, the sovereign in the United States is the embodiment of the people—their duly elected, representative government. The popularly elected legislature of each state is, in effect, both the sovereign and the public.[76] Thus, although American courts have consistently held that either the state or the legislature holds title to submerged navigable lands which are subject to the public trust, it appears that the legislature (or its delegate, typically a natural resource or coastal management agency) fills the roles of both the sovereign and the public. In traditional trust terms, the legislature acts as both the trustee and the representative of the beneficiaries of the trust, the public.[77]

3. Responsibilities and Obligations of the State as Trustee

Although many courts have ruled that the state holds public trust lands as a trustee, only a few decisions have addressed the actual duties and obligations of the state as trustee. Perhaps the most detailed and influential discussion of the duties and powers of the state as trustee is found in *Mono Lake*,[78] which quoted at length from *Illinois Central* and earlier California cases and concluded: "In summary, the foregoing cases amply demonstrate the continuing power of the state as administrator of the public trust, a power which extends to the revocation of previously granted rights or to the enforcement of the trust against lands long thought to be free of the trust."[79] However, despite its lengthy discussion of specific examples of trust rights, even the *Mono Lake* decision fails to explain or define the rights and obligations of the trustee in a coherent, comprehensive fashion.

4. Analogies to Private Trust Law

Given the consensus that states hold public trust lands in trust for the benefit of the public, and the dearth of cases which have addressed the obligations and

76. See *State v. Bleck*, 114 Wis.2d 454, 338 N.W.2d 492 (1983) (authority to administer the trust lies in the legislature).

77. For a discussion of the authority of a legislature to delegate its trustee obligations, see ch. 2, sec. D. See, e.g., *State ex. rel. Bd. of Univ. and School Lands v. Andrus*, 506 F. Supp. 619, 625 (D.N.D. 1981) (public trust lands are "held by the State of North Dakota in its sovereign capacity, and as trustee for the people of North Dakota"). See also ch. 1, sec. C, subsection 2 for discussion of establishment of the trust.

78. 33 Cal.3d 419, 189 Cal. Rptr. 346, 658 P.2d 709.

79. Id. at 440, 189 Cal. Rptr. at 360, 658 P.2d at 723.

responsibilities of the state as trustee, it is both reasonable and instructive for states and coastal managers to look to private and charitable trust law for guidance in determining their rights and obligations as trustees.[80] Analogies to private and charitable trust law suggest helpful approaches for managing public trust lands and resources and provide a framework within which to address difficult questions relating to the disposition of trust lands.

The fundamental premise in private and charitable trust law is that the trustee is in a fiduciary relationship to the beneficiaries of the trust—a relationship involving trust and confidence and the duty on the part of the fiduciary to act solely for the benefit of the other party as to matters within the scope of the relationship. In private trust law, the beneficiaries of the trust are readily identifiable as those persons set out in the trust instrument. However, in both charitable trusts and in the public trust context, the beneficiaries consist of a large group of people who are not specifically identified. Thus the trustee owes a fiduciary duty not to any specific person but to the community as a whole. The primary duties of private and charitable trustees to fulfill their fiduciary responsibilities are readily applicable to the public trustee. Each duty and the corresponding analogy to the public trust is addressed separately below.[81]

a. Duty of Loyalty. The most fundamental duty that a private trustee owes to the beneficiaries of the trust is the duty of loyalty. This duty requires that the trustee administer the trust solely in the interests of the beneficiaries and not act in its personal interest if such conduct might conflict with the interests of the beneficiaries. This is also known as the "duty of no self-dealing." Questions concerning a private trustee's duties of loyalty often arise when the trustee sells trust property to itself. Similarly, just as the trustee should not sell trust property to itself, it should not purchase property for itself that it should have purchased for the trust.

Furthermore, the trustee may not use the trust property for its own benefit without paying fair value for such use. Where the trustee does act for its own benefit, and where the beneficiaries have not been fully informed and given their consent, the transaction is voidable by the beneficiaries, even if the trustee acted in good faith and even if the transaction was otherwise fair and reasonable.

80. The United States Supreme Court has held that the "trust" obligations of the federal government with respect to federal lands are not on a par with a fiduciary's obligations under private trust law. *Alabama v. Texas*, 347 U.S. 272, 277 (1954). However, the federal lands "trust" derives from the express constitutional authority of the federal government over such lands (art. IV, sec. 2 of the United States Constitution) and is thus separate and completely different from the common law public trust doctrine. Therefore, courts and state legislatures and their delegates should not indiscriminately rely upon federal lands "trust" principles when confronted with issues involving the duty of the state as trustee under the public trust doctrine.

81. For a trustee's duties generally, see *Restatement (Second) of Trusts* 164, 169–85 (1959).

The duty of loyalty is easily adaptable as an analogy in the public trust context. The state as trustee has an obligation to exercise its rights and authorities over public trust lands and resources for the benefit of the beneficiaries—its people. If the state exercises its rights and authorities for the particular benefit of the state itself (e.g., selling public trust lands simply to raise revenues or balance its budget) or for the benefit of any one individual, or small group of individuals, that act could be challenged in most state courts as a violation of the state's duty of loyalty to the beneficiaries as a whole. To the extent that such a transaction is voidable under an analogy to private trust law, the question arises whether any one member of the public or group of citizens may bring an action to void a transaction based upon a violation of the state's duty of loyalty under the public trust. If the beneficiaries are the same as the trustee, i.e., the state, it may be that the state or an individual agency of the state has a right, as a representative of the beneficiaries, to bring an action to void any transaction that violates the state's duty of loyalty to its beneficiaries—for example, an action by the attorney general or the public trust agency to void any conveyance of public trust lands made in violation of the state's duty of loyalty to its beneficiaries.[82]

b. Duty Not to Delegate. A private trustee generally owes a duty to the beneficiaries of the trust not to delegate to others the entire administration of the trust or to delegate particular responsibilities that the trustee ought to perform. However, a private trustee is usually authorized to delegate certain responsibilities to the extent that the trustee would not surrender or lose supervisory control over the administration of the trust. For example, a private trustee typically is permitted to retain experts, such as lawyers, accountants, investment advisors, and others, but may not delegate essential functions or total responsibility for making legal or financial decisions or selecting investment opportunities to an outside party, and the trustee must retain and exercise ultimate supervisory and decision-making responsibility.

In the public trust context, the question of the duty not to delegate is analogous to the administrative law question of the legislature's authority to delegate its legislative functions.[83] Under basic administrative law, a legislature may not delegate its legislative functions to a state agency without in some way establishing standards for the delegatee agency to follow.[84] This rule essentially reflects the private trust analogy. For example, Wisconsin courts have addressed the issue of improper delegation of authority in the public trust context.[85] Although they did

82. See discussion on enforcement of the public trust doctrine, ch. 2, sec. E.
83. For a discussion of delegation generally, see ch. 2, sec. D.
84. Id.
85. In *State v. Village of Lake Delton,* 93 Wis.2d 78, 286 N.W.2d 622 (1979). The court held that the public trust may be delegated to other governmental units, including municipalities, for purposes

not make the specific analogy to private trust law, the Wisconsin cases raise the question whether the state may delegate its public trust responsibilities to either a state agency or municipality in such a way that the municipality or state agency has authority to convey public trust lands to private parties.

A related question is whether a state may delegate its responsibilities as trustee to private property owners. Could title be transferred to a private property owner with the private property owner thereafter responsible for administering the public trust? State laws or programs that license or authorize private owners to use trust lands for marinas, fisheries, and other trust uses effectively delegate day-to-day responsibilities for implementing and monitoring those uses to private parties. However, typically under such programs, the private owner is not permitted to change the use for which the initial license was granted without risking loss of the license.[86] Thus the state retains an oversight role protecting its trusteeship.

c. Duty to Furnish Information. A private trustee is responsible to provide the beneficiaries with complete and accurate information as to the administration of the trust. The duty to provide information raises specific problems where there are multiple beneficiaries. Should each member of the public have a right to obtain full information concerning all trust lands? Should only members of the public who have a specific separate interest in the trust lands have access to the information? How does the right to information relate to the public's rights to enforce the trust?[87] The problems created by the obligation to provide information to multiple beneficiaries may suggest that the state legislature or its delegate must act as both trustee and beneficiary. Thus the state legislature or its agency delegate must consider the interests of all its citizens in analyzing and reviewing information regarding public trust lands and resources. In addition to "freedom of information" and similar statutes that afford citizens access to public records in almost every state, many states in recent years have required that notice be published to the general public before any conveyance of public trust lands or

---

in furtherance of the public trust. The court upheld a municipal zoning-type statute restricting an area of a navigable lake for use solely by a private water ski promotion company. It held that the statute authorizing municipalities to regulate boating and other activities on navigable lakes was sufficiently specific to delegate authority for this type of regulation. This issue had also been recognized in *State v. Public Service Comm'n*, 275 Wis. 112, 81 N.W.2d 71 (1957), where the court found that delegation by the state to the city of Madison of the authority to build an auditorium and civic center on filled lakebed contained sufficient standards to govern the delegatee's conduct to authorize the delegation. The statute in question authorized the city to construct and maintain in or on the lake, but not beyond the "dock line," parks, playgrounds, bathing beaches, etc., and "public buildings." The new auditorium and civic center were found to be public buildings.

86. See, e.g., Massachusetts Chapter 91 program, discussed in chapter 5. See also *State of Vermont and City of Burlington v. Central Vermont Railway, Inc.*, 571 A.2d 1128 (Vt. 1989).

87. See ch. 2, sec. E for discussion of standing to enforce the public trust doctrine.

change in the use of such lands. Such notice should be provided as part of the trustee's duty to disclose.

   d. Duty to Take and Keep Control of Trust Property. A private trustee has a duty to take reasonable steps to establish and retain control of trust property. This duty ordinarily requires that the trustee take and keep possession of the property or specifically designate it as trust property (for example, by recording the trust interest in the appropriate registry of deeds).

   Under general public trust law, trust lands and resources may not be conveyed to private ownership except in certain circumstances.[88] The duty of the trustee to establish and retain control of trust property may further justify prohibiting states from conveying trust lands. Imposing an affirmative duty on the state as trustee to retain control over trust property for the benefit of the public (including the institution of proceedings to recapture public trust lands improperly conveyed to private interests) would provide members of the public additional grounds on which to challenge conveyances of trust lands and resources. Likewise, the affirmative duty on the part of the state as trustee to retain control over trust property may be a source of authority for state agencies to rely upon in determining their power to convey trust lands.

   e. Duty to Preserve the Trust Property and Duty to Deal Impartially with Beneficiaries. A private trustee has a duty to the beneficiaries of the trust to use care and skill to preserve the trust property. Where the trust property consists of financial assets, this obligation requires preservation of the principal. Where the trust property is real property, the obligation primarily concerns the maintenance, repair, improvement, and upkeep of the real property, including payment of taxes, mortgages, and fees.

   In the public trust context, the duty to preserve trust property might require considerably more diligence on the part of the public trustee in managing trust lands and resources than has yet been found necessary under the public trust doctrine. For example, the state could be required to adopt and enforce measures to provide environmental protection for trust lands and resources. Where public trust lands and resources are threatened by environmental degradation, pollution, or misuse, the trustee may have a duty to take action to protect and preserve those lands and resources.

   The duty to deal impartially with beneficiaries arises between successive beneficiaries. In the private trust context, there is often a conflict of interest between the current beneficiaries and future beneficiaries or remaindermen, who will receive the trust property in the future. The current beneficiary has an interest in

88. See ch. 2, sec. F for further discussions of the state's obligations when conveying public trust property to private ownership.

having the trustee invest the trust property to yield the highest current income. In contrast, the future beneficiary wants the trustee to exercise the utmost caution to preserve or expand the trust principal.

The private trustee is charged with the difficult task of managing the trust to benefit both current and future beneficiaries, while maintaining its impartiality. In the public trust context, the duty of impartiality toward beneficiaries raises a serious issue with respect to the state's obligation as trustee to preserve trust property for future beneficiaries, i.e., future generations. The analogy to the private trust duty to protect the interests of future beneficiaries, together with recent interest in the public trust doctrine as a means to preserve public trust land resources, provides the basis for an affirmative responsibility in the state to protect and preserve its public trust resources.[89]

A conflict for the state as trustee arises between the use of public trust property today and its preservation for the future. The public trust doctrine was not originally developed to protect trust resources, but was intended rather to protect the public's rights to have access and to use such resources. In particular, the public trust doctrine was intended to ensure that the general public had access to trust lands for the purpose of fishing, commerce, and navigation. Each of these uses involved use and exploitation of trust resources by current beneficiaries. Rarely, until recently, was thought given to the potential spoliation of such resources by present public trust use. The state's challenge today is to balance uses by current beneficiaries against the goal of preservation of trust resources for continued use by future generations.

Precedent exists in historical public trust law and more recent court cases to assist states in resolving this conflict. Under the traditional public trust doctrine, restrictions on exploitation (such as limiting the size of a catch) were seen as reasonable to protect everyone's access to the resources. Furthermore, California courts have held that preservation is a protected public trust use.[90] Thus, following this precedent and using the private trust analogy, states may claim authority to preserve public trust resources for use by future generations, even if preservation is challenged by proponents of what would otherwise be an appropriate public trust use. States have an obligation to balance present-day exploitation of

89. For example, in both *Marks* and the *Mono Lake* case, the Supreme Court of California held that the public trust doctrine affirmatively protects environmental and recreational values. Similarly in *United States v. State Water Resources Control Bd.*, 182 Cal. App. 3d 277, 227 Cal. Rptr. 161 (1986), the California Court of Appeals held that the public trust doctrine was an appropriate foundation for the promulgation of water quality standards that would protect fish and wildlife. The decision may be viewed as an endorsement of the state's use of the public trust doctrine to protect trust resources for both current and future generations.

90. See *Mono Lake,* 33 Cal.3d 419, 189 Cal. Rptr. 346, 658 P.2d 709; *Carstens,* 182 Cal. App. 3d 277, 227 Cal. Rptr. 135. See ch. 2, sec. B.

resources against future use and an obligation to regulate present-day exploitation to ensure the reasonable preservation of resources for future use.[91]

f. Duty to Enforce Claims. A private trustee has a duty to its beneficiaries to enforce claims related to the trust property. For example, a private trustee may be liable if it delays in bringing a lawsuit and the party obligated to the trust or the beneficiaries has become insolvent. A private trustee may also have a duty to bring an appeal if it should reasonably do so under the circumstances.

The duty to enforce claims raises the possibility in the public trust context that there is an affirmative obligation on the states to protect public trust interests in lands already in private ownership.[92] States may be compelled to reassert public trust interests in lands long held by private parties. Where a state does not act to enforce the trust, the attorney general or private parties may be entitled to compel enforcement.[93]

g. Duty to Make Trust Property Productive. A private trustee has a duty to invest trust funds so that they will be reasonably productive of income.[94] Where the trust estate consists of real property, "productivity" often depends on the terms of the trust and the surrounding circumstances. For example, it may be the duty of the trustee to sell the land and invest the proceeds. Where the trustee is not directed or empowered to sell the real property, it generally has the power to lease the land in order to make it productive and to generate income. Alternatively, in the case of an agricultural property or otherwise productive property, the terms of the trust might authorize the trustee to manage the land and generate income from its produce.

The duty to make trust property productive might initially appear to raise difficult issues in the public trust context, since in the private trust context, productivity is typically measured by income generation. However, in the public trust context there is a serious issue whether productivity should be measured by income generation—i.e., development potential—rather than by the benefits to the public as beneficiaries—i.e., maximizing the availability of public trust lands

91. It must be remembered, however, that a state's legitimate interest in conserving natural resources which it holds in trust for all its citizens cannot be exercised in a way that unreasonably burdens interstate commerce or unfairly discriminates against citizens of other states. See, e.g., *Hughes v. Oklahoma,* 441 U.S. 322 (1979).

92. For example, does a state have an obligation, as the state did in *Illinois Central,* 146 U.S. 387, to challenge past conveyances of trust lands for private purposes?

93. See discussion of enforcement of the public trust doctrine, below, ch. 2, sec. E.

94. This section does not suggest that every element of private trust law can or should be incorporated into the public trust doctrine. Obviously certain elements appear more compatible with the public trust doctrine than others. Nevertheless, these latter elements are discussed in order to demonstrate how the issues they raise (e.g., the duty to make trust property productive) may be addressed in the public trust context.

for public trust uses. Given the modern trend toward preservation of natural resources, states can and should make a strong argument that "productivity" in the case of the public trust may be served by enhancing public access to and preserving trust resources for current and future generations.[95]

Another requirement in both private and charitable trust law is the duty of the trustee to diversify the investments or holdings of the trust, in order to minimize risk. Diversification in the private or charitable trust context typically means diversification of investment instruments. For example, a trustee may be required to invest not more than a reasonable percentage of the trust assets in any one investment instrument, such as common stocks, corporate bonds, and governmental obligations, unless specifically instructed to do so under the terms of the trust.

The diversification requirement may not be directly applicable in the public trust context. However, the analogy may be used by states to support a decision by the legislature or its delegate to diversify the state's holdings in public trust land so as not to risk their destruction and the resultant inability of future generations to enjoy their use. This duty may imply an obligation of the state, as trustee, to maintain certain public trust lands and resources in a natural state for preservation, while developing other lands and resources for public uses.

5. Analogies to Charitable Trust Law

Charitable trusts are governed by essentially the same trust law as private trusts. However, special rules have been developed to address two important differences. The most obvious is that the beneficiaries of the charitable trust are numerous and not easily identifiable. By comparison, the beneficiaries of a private trust are typically listed in the trust instrument and are easily identified. The other key difference between charitable and private trusts is that charitable trusts may be unlimited in duration, as long as they are designed to accomplish goals that are beneficial to the community. Private trusts, by contrast, must in most states be of limited and defined duration, under an ancient property law doctrine known as the "rule against perpetuities" or its modern statutory equivalent.

Perhaps the most interesting special rule of charitable trust law with implica-

95. Alternatively, states may argue in certain circumstances that the most productive use of public trust lands is by development that opens access that is otherwise absent or limited, such as development of public marinas or public parks on dangerously decaying piers, on previously filled land used as a refuse dump, or on isolated and otherwise inaccessible beach or island areas. It is not inconsistent with either public trust or environmental principles to recognize that in some circumstances the public interest can best be served by advancing private interests, because the public derives substantial benefits secondarily [e.g., *Fallbrook Irrigation District v. Bradley*, 164 U.S. 112, 161 (1896)]; or because the particular kind of private interest is so widely shared as to make it a prime object of governmental interest [cf. *Pennsylvania Coal Co. v. Mahon*, 260 U.S. 393, 413 (1922)].

tions for the public trust doctrine is the *cy pres* doctrine. Under this doctrine, a court can change the terms of the trust in appropriate circumstances. Specifically, if the trustee is able to show that the purpose for which the charitable trust was created is either illegal or effectively impossible or substantially impracticable to carry out, courts have the authority to permit the trust to be applied to another related or similar purpose that would accomplish the general goals established under the trust.[96]

Under British law, the authority to change the terms of a charitable trust lies with the Parliament.[97] In the United States, however, legislatures are without authority to change gifts already made—this authority lies in the courts. The *cy pres* doctrine may be a tool to be used by states in expanding the range of public trust uses. As described earlier, the original uses under the public trust doctrine were fishing, commerce, and navigation, although many courts have expanded the list of uses over time. Courts have relied upon a variety of legal theories to increase the number of public trust uses.[98] However, further expansion might be better grounded in an analogy to the *cy pres* doctrine. The reasoning is analogous—the historical purposes of the public trust may, in particular circumstances, have become outdated or the uses traditionally protected by the trust no longer fully serve the purposes for which the public trust doctrine was created. Thus, in a manner similar to the use of the *cy pres* doctrine, courts might well be convinced to allow the state as trustee to permit public trust lands and resources to be used for a broader spectrum of water-dependent or water-oriented uses reflecting the needs and interests of the contemporary public.

Under general trust law, the terms of the trust may be enforced either by a beneficiary or by one of the trustees. Charitable trusts, as stated above, usually have no specified beneficiaries. Therefore, a rule of law has been established giving the state attorney general the right to enforce charitable trusts as a representative of all beneficiaries. In addition, there are two other ways to enforce charitable trusts. First, an individual beneficiary may encourage the attorney general to bring an action where the individual is the "relator." In this situation, the attorney general brings the lawsuit based on a specific claim or complaint of a

96. An example of the exercise of the *cy pres* doctrine involves a trust for the development of a vaccine for polio. Once a polio vaccine had been discovered, the *cy pres* doctrine might be available to alter the terms of the trust to channel its resources into the development of vaccines for other diseases, or into the treatment of polio other than by the development of the vaccine.

97. For example, many early British charitable trusts were established for the benefit of grammar schools to teach Latin and Greek. In the early twentieth century, masters of the grammar schools wanted to expand the curriculum beyond Latin and Greek. The British courts refused to change the terms of the trust. However, the Parliament then passed statutes that enabled changes in the curriculum.

98. See ch. 2, sec. B.

beneficiary. Second, any beneficiary with a "special interest" in the performance of the charitable trust has the right to bring a lawsuit to enforce it. However, such an individual must demonstrate that the interest is special in that it derives from some benefit under the trust that is different from the benefit generally available to all beneficiaries. Furthermore, even in such special interest cases, the attorney general is usually a necessary party to the action.

In the public trust context, enforcement of the trust becomes somewhat complex.[99] If the attorney general is the party to bring the lawsuit on behalf of all beneficiaries and the lawsuit is brought against the trustee, then the attorney general would effectively be suing the state or its delegate, the administrative agency.[100] Thus enforcement is more likely to be limited as a practical matter (unless a statute dictates otherwise) to beneficiaries who have "special interests" and are willing to bring an action against the state or agency. For example, a fisherman who can prove that he has consistently used a public dock over a very long period of time might have a special interest that would allow him to bring a lawsuit against a state agency which permitted the licensing of the dock for private use only or for "dockominiums."[101] A limitation identified by this analysis is that of cost. Where the only enforcement mechanism available is a lawsuit brought by a private party with a "special interest," and where a court cannot award monetary damages, there may be little incentive for private parties to bring costly lawsuits to enforce the public trust—at least without the incentive of a statutory entitlement to an award of legal fees for the prevailing party.

6. Conclusion

The analogies to private and charitable trust law may be fruitful sources of legal and practical approaches to implementation of the public trust. However, the question remains whether courts could be persuaded to hold the state as a public trustee to the fiduciary responsibilities of a private or charitable trustee.[102] The likely case to test this issue would involve a lawsuit by a private citizen as a special interest beneficiary of the trust against a state agency for breach of the state's fiduciary obligations. To date, few cases have been brought on a straight trust/

99. See generally ch. 2, sec. E.

100. See, e.g., *Byrd v. Mississippi*, Ms. Chanc. Ct., Harrison Cty., 2d Dist., No. 17, 879 (April 18, 1990), where the Mississippi attorney general sued the state, the city, and the county to enforce the public trust doctrine and for violation of the state constitutional provision barring the state from making gifts without explicit legislative authority in the face of a statute setting the public trust boundary at the 1973 mean high watermark. The court upheld the statute.

101. See discussion of marinas and harbors, ch. 3, sec. C, subsection 7.

102. But see *Slocum v. Borough of Belmar*, 238 N.J. Super 179, 569 A.2d 312 (1989), where the court held that a municipality had the responsibilities of a trustee, using analogies to private trust law.

fiduciary obligations theory. Thus it remains to be seen how successful such a theory could be.

Pending such cases, however, states and coastal managers may affirmatively use the private and charitable trust analogies to strengthen their control and enforcement over public trust lands. For example, states and their delegates may clothe themselves in the mantle of a trustee, in the traditional private and charitable trust sense, in order to assert greater control over the management and preservation of public trust resources for both current and future generations by explicitly incorporating private and charitable trust principles into regulations, using guidelines and other regulatory activity affecting public trust lands and resources.

D. LEGISLATIVE DELEGATION OF PUBLIC TRUST RESPONSIBILITIES

This section addresses the question what persons or agencies within the state have the authority to implement the public trust doctrine. Authority to act under the public trust doctrine must be delegated from the state legislature to a state agency or official. This section provides a brief background on the principles of legislative delegation in general and their relationship to delegation in the public trust context. It then describes the potential for challenge to the authority of state coastal agencies based upon an improper delegation theory and examines ways to respond to such challenges. Finally, this section discusses the significance of agency rulemaking and procedures in upholding the agency's authority to act under the public trust doctrine.

What person or agency has authority to act to enforce the public trust? In many coastal states, a statute expressly authorizes a particular state agency to exercise full authority over the lands and waters in the coastal zone through comprehensive coastal zone management legislation, sometimes explicitly referencing the public trust. In others, existing statutes, such as clean water or wetlands protection laws, have provided the basis for an agency to establish coastal management programs, including public trust authority, without explicitly mentioning the public trust. In some states, courts have interpreted existing legislation to confer public trust authority upon an agency already vested with power over some aspect of coastal development or activities within tidelands. Some coastal agencies, on the other hand, have been given, either by the terms of their enabling legislation or as a result of judicial decision, relatively narrow authority restricted to granting permits and licenses or to strictly conservationist or protectionist activities, rather than broad public trust responsibilities or general power over coastal development management and control programs. In order to be in a position to exercise their authority as intelligently, effectively, and efficiently as possible, coastal managers in each state must know the scope as well as the source of their jurisdiction and powers, including those relating to protection of the public trust. This is particularly important in light of the fact that their coastal policies, programs, and decisions will inevitably affect the economic and property interests of people and organizations who will frequently challenge their efforts legally.[103]

Because of the peculiar political history of the United States, one of the most basic and, at least until relatively recently, most common challenges to administrative agencies, including coastal management agencies, has been based upon the theory of improper delegation of authority by the legislature to the agency.[104] Influenced in part by their own experiences with the perceived abuses of authority by the English king and Parliament, the colonists established state and federal constitutions that created governments that, continuing to this day, are characterized by structural checks and balances. Such checks and balances theoretically prevent any one branch of government from attaining absolute power or dominating the other branches or the citizenry generally. The principal mechanism for this constitutional system is the allocation of governmental powers among three branches—the legislative, executive, and judicial—each of which, under the doctrine of separation of powers embedded in the federal and state constitutions,

103. In some states (e.g., California), there may be an effective statutory division of public trust responsibilities, as, for example, between an agency with authority over coastal development including tidelands and another with power over activities in submerged lands seaward of the low watermark; or between the agency entrusted by statute with jurisdiction over the use and disposition of state-owned lands and the agency charged with regulating activities along the coast but without property-disposition powers. It is important for each coastal agency to identify all other agencies that may "hold" or "exercise" the public trust, if only for purposes of an efficient division of labor in the handling of their regulatory workloads. Because of the functional similarity of regulatory licenses and leases, the problems of duplication of effort and potential inconsistency in applying public trust principles by "property" agencies and "licensing" agencies are particularly likely to arise and should be consciously addressed. In any event, as in the regular law of trusts, all agencies charged by statute with any aspect of responsibility for the trust are the equivalent of "co-trustees," with each having a separate and nondelegable obligation to administer the trust, within the scope of its delegated authority, independent of the activity or inactivity of the others. In the absence of statutory guidance, it is not merely rational but essential for such agencies to work together and coordinate their endeavors if the public trust is to be adequately vindicated and protected. See generally the discussion of licensing in ch. 2, sec. G.

104. Compare, for example, *Adams v. North Carolina Dep't of Natural and Economic Resources*, 295 N.C. 683, 249 S.E.2d 402 (1978), upholding the North Carolina Coastal Area Management Act of 1974 against an attack asserting that the statutory charge to the agency to develop and adopt guidelines for the coastal area was an unconstitutional delegation of authority, with *Askew v. Cross Key Waterways*, 372 So.2d 913 (Fla. 1978), holding unconstitutional provisions of the Florida Environmental Land and Water Management Act of 1972 delegating to the Florida Division of State Planning the power to designate areas of critical state concern within which development would be controlled.

can exercise only its own special powers. They cannot, in strict theory, either exercise the powers of the other branches or delegate their special powers to the other branches or to any other agencies of government, or to private persons.

Thus the legislative power to make discretionary choices of public policy and to promulgate general rules of conduct by enacting legislation embodying the chosen policies—i.e., to make law—was supposed to be exercised only by the legislature and was not to be delegated by it; while the judicial power—to hear cases and controversies and make decisions concerning private and public rights, interests, and obligations—was similarly not to be delegated beyond the courts. A century later, however, it had become clear that the practical necessities of modern government were incompatible with rigid adherence to the literal theories of separation of powers and nondelegation. The ever-increasing need for continuous expert supervision and systematic, uniform regulation of a host of areas of socioeconomic concern in a complex and dynamic society gave birth (in the late nineteenth and early twentieth centuries) to administrative agencies, whose essential characteristic—the combination of legislative and judicial powers in an agency within the executive branch—has required modification of the strict nondelegation doctrine.

Currently, in virtually every state and at the federal level, delegation of even extensive legislative and quasi-judicial powers to an administrative agency is almost always upheld against a constitutional challenge under the separation of powers or nondelegation theory, provided three conditions are satisfied. First, the agency must be acting in accordance with its enabling act, the statute that creates the agency or confers upon it authority with respect to a particular subject matter. The enabling act is the fundamental source of the agency's power, which both defines what the agency may do and limits the extent of that authority. Agency actions within the scope of the statutory authority expressly or impliedly conferred upon it by the enabling act are presumptively valid exercises of delegated powers; actions outside or beyond the scope of that authority are invalid, sometimes under the label of *ultra vires*. Consequently, it is essential for coastal managers in each state, with the assistance of their legal advisors, to be fully familiar with the provisions of their enabling act, as well as with all judicial and prior agency decisions interpreting or applying it and all general judicial restrictions upon the exercise of power by administrative agencies in the state generally.

Second, the enabling act or empowering statute must provide meaningful standards setting forth the legislatively determined policy goals or objectives that are to guide the agency in the exercise of its delegated discretion. Without statutory standards, agency action will not pass constitutional muster, since it would be nothing more than the impermissible exercise of unfettered (and judicially unreviewable) discretion to engage in arbitrary lawmaking. So long as some standards are provided by the legislature, the agency is free to administer, fill up, or work out the details of the statutory policy without doing violence to the separation of powers principle. In earlier years, courts commonly insisted that the enabling statute contains express or detailed standards. However, many statutory standards for agencies have been upheld despite being far from explicit or helpful.[105] Since harried legislatures usually fail to include any but the broadest and most general standards in their enactments, modern courts are typically willing to tolerate and constitutionally approve delegation statutes even in the absence of explicit standards, provided the policy and purpose of the legislature are either expressed or can be implied so that they can serve as the measuring rod for agency action, which must have a rational relationship to such policy and purpose to be valid.[106]

Finally, even when statutory standards or policies are less than clear, most state and federal courts are willing to uphold legislative delegations to agencies, provided that the agency's exercises of authority are: subject to reasonable procedural safeguards imposed either by the legislature (in the enabling act or an administrative procedure act) or adopted by the agency itself in the form of procedural regulations; in conformity with meaningful substantive standards promulgated in agency regulations with the stated purpose of implementing statutory goals; and ultimately subject to judicial review. Thus, even when the legislature has not given specific directions, agencies should be able to deflect a delegation challenge by pointing to their own enunciation of reasonable and intelligible standards rationally related to statutory policy and their own provision of fair procedures for those affected by their actions.[107]

---

105. For example, observe such broad phrases as "just and reasonable rates" in the Interstate Commerce Commission's railroad regulation; "public interest, convenience, or necessity" in the Federal Communications Commission's issuing of radio broadcast licenses; and "unfair methods of competition" in the Federal Trade Commission's regulation of business conduct, all of which have been judicially upheld as legally adequate.

106. One of the principal reasons courts are willing to imply the existence of standards is the doctrine, recognized in every jurisdiction, that courts are obligated to interpret a statute, if at all possible, so as to avoid serious doubts as to its constitutional validity.

107. To a great extent, an agency can achieve most if not all of its purposes by conscientiously following the principles of fairness and common sense—fairness in the methodology by which decisions are reached; fairness in the decisions reached; and the use of sound scientific and technical procedures and valid data to support its assumptions and conclusions. It should be noted that state courts, at least until relatively recently, have tended to be somewhat more restrictive in their separation of powers decisions than federal courts and have often disapproved the delegation of lawmaking powers to state agencies if the enabling act lacks what the court thinks are meaningful standards. This seems to have occurred most frequently when the delegated power involved the licensing of a trade or profession or the private use of real property (especially zoning). Court opinions ostensibly based on the lack of meaningful standards are often actually premised on the courts' concern that the delega-

Several specific issues in the area of delegation are worthy of mention. First, probably the most important power delegated to administrative agencies is the authority to promulgate rules or regulations, since rule making has come to occupy a central position in modern administrative law as the most effective means for implementing the legislature's broad policy statements. Like all agency powers, rule-making authority must be derived either from the express provisions of the enabling statute creating or empowering the agency or by necessary implication as reasonably required for the agency to effectuate the purposes of the legislature as reflected in the terms of the statute. Most modern statutes do confer rule-making authority on agencies in broad and express terms, since the very rationale of the modern administrative agency lies in the recognition that legislatures are unable or unwilling to provide detailed, experienced, and ongoing regulation of essential areas of governmental concern.

Yet most courts will uphold agency regulations, even when the authority cannot be traced to specific words in a statute, whenever they conclude that the exercise of regulatory power is necessary to carry out the purposes expressed in the statute and whenever they can find at least some rational relationship between the regulation and the purposes of the empowering statute. The implication of authority cannot, however, arise from a total statutory vacuum, and courts will not uphold regulations that have no relationship to, or are inconsistent with, the statutes creating the agency. Therefore, an agency's ability to cite a legislative or judicial basis for its rule making, or to support it as in furtherance of statutory goals, is of great importance.

Valid agency rule making is essential to a coherent coastal management program for several reasons. Valid agency regulations are given the same legal force and effect as a statute and are accorded the same presumptions of validity as a statute. Furthermore, an agency's reasonable interpretation of its own regulations will be accorded great deference by the courts. Finally, challenged regulations will be upheld as lawful exercises of discretion as long as they are supportable on any rational basis, placing the burden on those opposing them to prove the absence of any conceivable ground on which they can be upheld.[108] On the

---

tions have been made to biased officials, involve an individual's livelihood or most important asset, may not require procedural safeguards, or may be susceptible to discriminatory administration. Although the pattern among the states is uneven, it is unlikely that any coastal management agency which has promulgated reasonable substantive standards expressly stated as being in furtherance of statutory policy and which utilizes fair procedural safeguards to govern its actions would have to be concerned about delegation attacks.

108. Agencies must, of course, follow the procedures specified for the promulgation of regulations in the enabling act or the applicable state administrative procedure act or their own regulations for their rules and regulations to be presumptively valid. Furthermore, under the *ultra vires* principle, rules and regulations can extend no further than the scope of the authority expressly or by necessary

basis of these principles, any agency that has any statutory jurisdiction over tidelands or coastal areas should have little difficulty in implementing public trust principles by rule making and regulation.[109]

Finally, certain other delegation issues should be mentioned: (1) the power of eminent domain—the power to take private property for public use—is a legislative power that may properly be delegated to an agency, so long as fair procedures are provided for determination of the issue of just compensation for the taking, and the issue of whether the exercise of the power is for a public purpose remains subject to judicial review; (2) the power to impose a civil penalty for violation of regulations may be delegated to an agency where it is reasonably necessary to assist the agency in the performance of its duties, where a reasonable upper monetary limit is set by the legislature, and where the purpose of such penalties is not to punish the commission of a crime through the disguised device of a civil

---

implication conferred on the agency by the enabling statute. Thus an agency specifically authorized to issue permits for wharfing out or for coastal dredging and filling activities will be empowered to make its permitting decisions (including disapproval of applications even if they meet statutory or regulatory performance standards) in light of their impact on interests protected by the public trust doctrine, but may not (in the absence of expansive judicial interpretation of its enabling act) be broadly authorized to develop regulatory guidelines or land use plans for all uses of coastal areas under the public trust doctrine. In any event, because of the presumptive and practical validity of properly promulgated regulations, coastal management would be wise, particularly in states where the source of their public trust authority is neither explicit nor clear, to incorporate, consciously and deliberately, public trust principles into the text of their regulations and also to articulate expressly therein the promotion or protection of the public trust as the basis and reason for the regulations, with as much reference to their enabling act as circumstances allow. See ch. 3.

109. In view of the extremely broad discretion possessed by a coastal agency—not only because of these deferential principles of administrative law applied by the courts but also because coastal agencies are not merely regulating under the general police power but are in fact regulating the use of public property (i.e., the lands subject to the public trust)—it is likely that virtually any rule or regulation promulgated by the agency that is explicitly and thoughtfully based upon protecting or advancing the public trust will be upheld against challenge, so long as the agency fully articulates the regulation's basis and purpose, including an identification of the public trust interests involved. Consequently, a coastal agency selectively and judiciously applying public trust principles could in all probability create valid regulations (including regulations classifying types of shores and coasts) which, for example, preserve particularly sensitive shorelines in their natural state and prohibit any development thereon; create temporary moratoria or freezes on all development, or on certain types of development; prohibit activities with unduly adverse impact on water quality; give aesthetic values and considerations (related to enjoyment of the coastal area) primacy in evaluating projects; or authorize specified public trust uses in certain areas to the exclusion of others (so long as reasonable opportunities to engage in the others are not totally extinguished). In short, coastal agencies applying public trust principles should certainly possess at least the breadth of power acknowledged in contemporary zoning, land use, and resource allocation agencies [see, e.g., *Berman v. Parker*, 348 U.S. 26 (1954); *Citizens to Preserve Overton Park v. Volpe*, 401 U.S. 402 (1971); *Almquist v. Town of Marshan*, 308 Minn. 52, 245 N.W.2d 819 (1976)]; and undoubtedly will be granted even greater flexibility and leeway by the courts because of the public proprietary nature of the areas subject to regulation—assuming, as always, consistency with the agency's enabling statute and a bona fide relationship to public trust interests.

penalty; (3) an agency that has (or whose head has) been delegated rule-making and adjudicatory powers may not, in the absence of explicit statutory authorization, itself delegate (i.e., subdelegate) discretionary (as opposed to purely routine or ministerial) actions either to subordinate agency officials or to other governmental units (e.g., municipal authorities) or to groups of private persons; and (4) the power to exact license fees may properly be exercised by an agency, even in the absence of explicitly statutory language, as long as the agency has been clearly delegated licensing powers and the fees are rationally related to the agency's licensing activities.[110]

E. ENFORCEMENT OF THE PUBLIC TRUST

This section addresses situations where either a state agency, official, or citizen seeks to enforce the public trust doctrine. It first considers who among the public has the right to bring an enforcement action and under what legal theories such a person could bring an action. In the same context, this section provides a brief description of the role of the state attorney general, who may be called upon to act on behalf of the public or a specific member of the public in order to enforce the public trust doctrine against a private owner or state agency. This section then considers the potential for enforcement of the public trust doctrine by state coastal agencies against private property owners or other state agencies or officials.

How is the public trust enforced? The public trust rights that have been identified and described in the preceding sections are not self-executing and exist primarily because courts have enforced them. Without judicial recognition there would be no public trust doctrine. Along with these judicially created rights comes the vehicle by which they can be exercised and elaborated—the availability of a judicial remedy. Coastal managers and other state officials must interpret and apply existing judicial pronouncements concerning the public trust in devising and implementing enlightened public policy. However, the legal validity of such policies, and ultimately the breadth and scope of the public trust doctrine itself, will be determined by the courts.[111]

---

110. For a more detailed discussion of licensing, see ch. 2, sec. G. See generally on the subject of delegation, K. C. Davis, *Administrative Law Treatise*, vol. 1, ch. 2 (2d ed. 1978).

111. Given the recent withdrawal of the United States Supreme Court's mandatory appellate jurisdiction over state court decisions where a state statute is upheld against constitutional challenge, see Pub. L. No. 100–352, 7 102 Stat. 662, 664 (1988) (amending 28 U.S.C. 1257), and the states' immunity from suit in a federal court under the Eleventh Amendment of the United States Constitution, the future of the public trust doctrine will be entrusted in large measure to the courts of the several states and the lower federal courts. As a matter of both sound public policy and legal precedent, the fate of the public trust doctrine is in any event placed in the state courts. Absent

Coastal managers must be prepared to defend suits claiming that their actions regarding coastal or tidal property violate the public trust doctrine.[112] Individual managers may be named as defendants in such a lawsuit.[113] Most such lawsuits seek prospective injunctive relief—asking the court to order the agency to take or not to take some action in the future. However, if money damages are sought, individual defendants should consult the laws of their state to determine the state's policy toward indemnifying them against a judgment rendered against them.[114]

Lawsuits to enforce the public trust doctrine are most often brought by a member of the general public who disagrees with the use to which public trust lands or resources have been or are to be used. In defending such a claim, the agency must first determine, as a procedural matter, whether the person bringing the suit has standing to bring the lawsuit—that is "whether the litigant is entitled to have the court decide the merits of the dispute or of particular issues."[115] In the absence of a special state constitutional provision or statute conferring standing on citizens with no direct interest in the matter to sue to enforce environmental rights or a judicially recognized right of citizens with no direct interest in the matter to assert as a "private attorney general," the rights of the public as a whole on matters including "public rights,"[116] the person bringing the suit must establish his or her standing, or right, to sue a governmental agency. The generally accepted rule for determining whether a private litigant has standing to bring

---

interference with navigation, interstate commerce, or the federal government's defense needs, it is not apparent that anything in the Constitution or federal law authorizes a federal court to oversee or limit a state's determination concerning the public trust. As discussed elsewhere in this book, federal courts have acted, on various grounds, to "police" state implementation of the public trust doctrine. These cases have usually concerned allegedly improper conveyances of trust lands and termination of public rights. See ch. 2, sec. F and ch. 4.

112. Suits by concerned citizens and environmental groups attacking state and federal agency decisions affecting the use of natural resources have multiplied in recent years. Whether such an action can name a given state agency as the defendant or whether the defendant will be the state itself will largely be determined by whether the agency's enabling statute grants it the right to sue and be sued. It should be noted that such a general statutory grant is insufficient to waive a state's sovereign immunity under the Eleventh Amendment from suit in a federal court. *Attascadero State Hosp. v. Scanlon,* 473 U.S. 234 (1985).

113. Naming individual managers as defendants rather than the state or its agency directly avoids the Eleventh Amendment bar to suing a state in federal court. *Ex parte Young,* 209 U.S. 123 (1908). If the suit is for retroactive money damages, however, which must be paid out of the state's treasury, as opposed to prospective injunctive relief, the Eleventh Amendment bar still holds. *Edelman v. Jordan,* 415 U.S. 651 (1974).

114. Typically, individual state officials will either be legally immune from or indemnified for money judgments that arise out of the good faith performance of their authorized duties.

115. *Allen v. Wright,* 468 U.S. 737, 751 (1984) [quoting *Warth v. Seldin,* 422 U.S. 490, 498 (1975)].

116. *See* discussion in ch. 3, sec. A. See, e.g., *Marks v. Whitney,* 6 Cal.3d 251, 261–62, 98 Cal. Rptr. 790, 491 P.2d 374 (1971); *Brewster v. Sherman,* 195 Mass. 222, 80 N.E. 821 (1907).

such a suit is: can the litigant demonstrate that the action he or she challenges has created or will create an injury in fact and that he or she will be among the injured?[117] This requirement of a particularized injury in fact is intended "to put the decision as to whether review will be sought in the hands of those who have a direct stake in the outcome," and to insulate governmental agencies from attacks by "organizations or individuals who seek to do no more than vindicate their own value preferences through the judicial process."[118]

Such thinking has led courts to conclude that, absent a litigant with an injury in fact, a public right is most appropriately vindicated by the arm of the government charged with enforcing such rights—usually the state's attorney general (but also a state agency or public officer, such as a coastal management agency, if a state statute expressly so authorizes).[119] This view is supported by the traditional perception that where a state has entrusted the vindication of certain public rights to duly empowered state officials, the discretion of such officials should be relied upon and not preempted by piecemeal and uncoordinated intervention by members of the public.[120] Some courts, however, have determined that reliance on the heavily overburdened resources of an attorney general will not adequately protect and enhance the environment. These courts have constructed and deployed what amounts to a private attorney general theory, which permits a member of the public to sue to enforce a public right. Such a

117. This requirement of individualized injury was elaborated by the United States Supreme Court in *Sierra Club v. Morton*, 405 U.S. 727 (1972), where the Sierra Club, a nationally renowned conservation group, attempted to restrain a federal official from approving an extensive ski area in the Sequoia National Forest. The Court held that the club's special interest in preserving the forest was insufficient to give it standing. The Court went on to deny standing because "[t]he Sierra Club failed to allege that it or any of its members would be affected in their activities or pastimes by the Disney development." 405 U.S. at 735. The *Sierra Club* standard was relaxed somewhat in its later interpretations in *United States v. S.C.R.A.P.*, 412 U.S. 669 (1973) and *Duke Power Co. v. Carolina Environmental Study Group*, 438 U.S. 59 (1978), the effect of which was to give standing to environmental groups that can demonstrate actual injury in fact to specific members of the group. The decisions in *Sierra Club*, *S.C.R.A.P.*, and *Duke Power* are interpretations of the "cases and controversies" provision of Article III of the Constitution, which determines the jurisdiction of the federal courts. It is not directly applicable or binding on state courts, many of which are not restricted by the "case or controversy" standard. Nevertheless, many state courts have embraced its reasoning.

118. *Sierra Club*, 405 U.S. at 740. One should note that in most states private individuals may participate as intervenors in an adjudicatory proceeding before a state agency merely by being an interested person, with the right to present arguments orally and in writing, even if such persons do not have standing to be a full-fledged party because they are unable to show they may be substantially and specifically affected by the proceedings. E.g., Mass. Gen. L. ch. 30A, sec. 10 (1978).

119. E.g., *Nugent v. Vallone*, 91 R.I. 145, 161 A.2d 802 (1960).

120. *Cort v. Ash*, 422 U.S. 66 (1975); see also *Scott v. Chicago Park District*, 66 Ill.2d 65, 360 N.E.2d 773 (1977), upholding the Illinois attorney general's right to bring an action to enforce the public trust doctrine after the same action had been dismissed when brought by private citizen in *Droste v. Kerner*, 34 Ill.2d 495, 217 N.E.2d 73 (1966).

private citizen must demonstrate that he will suffer a specific injury other than that which is suffered by the public generally.[121]

To guard against the danger of multiple or frivolous lawsuits, some courts have required that a private plaintiff in such cases must bring a class action.[122] Other courts have held that any taxpayer and citizen has the right to seek judicial redress on matters of significant public concern so long as there is no more directly affected individual who is likely to bring suit.[123] No court to date, however, has gone the next step to award attorneys' fees and costs to a private plaintiff asserting public rights in lieu of the state's attorney general.[124]

The question whether a citizen has standing to enforce the public trust doctrine might lose most of its significance and complexity if the protected public interest were characterized as a traditional vested property right. However, while some movement in this direction occurred in the nineteenth century,[125] modern jurisprudence has tended to ignore it, and the few courts that have addressed the issue have held that a citizen may not avoid the injury in fact standing requirement by claiming a property interest in public trust land.[126]

Assuming a citizen succeeds in establishing standing to bring a lawsuit to protect or vindicate the public trust, the citizen still has the burden to state a claim for which relief can be granted—that is, to articulate a legal theory under which the court is empowered to grant relief. Because the public trust doctrine does not give any individual an identifiable and vested property right capable of recordation (as well as the fact that the property rights at issue are the public's generally and not the individual's), the litigant loses any potential claim under

---

121. See *Bedford v. City of New Haven,* 170 Conn. 46, 364 A.2d 194 (1975); *Marks,* 6 Cal.3d 251, 98 Cal. Rptr. 790, 491 P.2d 374 (permitting member of the public to determine existence and extent of public easement in tidelands). Excellent examples of specially affected individuals who should have standing to sue to vindicate public trust interests would be those who rely on access to the shore and the tidelands for their livelihood, such as fishermen, shellfish harvesters, fish processors, and those engaged in boat servicing and repair.

122. E.g., *Akan v. Olohana Corp.,* 65 Haw. 383, 652 P.2d 1130 (1982). A class action is a procedural device that has become increasingly employed against agencies and officials in which some member of a class of persons alleged to be commonly and adversely affected by official action is allowed to sue on behalf of the entire class, many of whom might not personally be impacted sufficiently to have the incentive to sue or may be unable to do so, provided that the plaintiff is able to demonstrate it will be a typical representative who will adequately and fairly protect the interests of the class.

123. E.g., *Trustees for Alaska v. State,* 736 P.2d 324 (Alaska 1987).

124. But see, e.g., Cal. Code Civ. Proc. 1021.5, which authorizes an award of attorneys' fees to private litigants who bring actions that result in public gain.

125. The United States Supreme Court gave support to this property right notion in *McCready v. Virginia,* 94 U.S. 391, 395 (1876). The Court there defined the public trust as a property right inhering in each individual citizen of a state as distinguished from a mere privilege or immunity of citizenship.

126. E.g., *Kerpelman v. Board of Public Works of Maryland,* 261 Md. 436, 276 A.2d 56 (1971).

the state or federal constitution that his property has been "taken" by the government and that he is entitled to just compensation.[127] The litigant should similarly be prohibited from claiming that the property has been taken without due process of law in violation of the Fourteenth Amendment. Therefore, if a state or state agency is sued on the grounds that official action regarding public trust lands has deprived members of the public of constitutionally protected property rights, the state or agency should move to dismiss the action.[128]

In certain circumstances, an individual litigant might have a sustainable claim for improper delegation of authority or for acts which are *ultra vires*.[129] Legislatures delegate authority to implement their directives to administrative agencies such as coastal commissions. Despite generally liberal attitudes toward most exercises of agency power, courts have tended to take a narrow view of a legislature's delegation of authority in connection with the alienation of public trust lands, and such decisions made by nonelected agencies rather than by the legislature itself will be subjected to closer scrutiny than will legislative decision making.[130] Some courts have gone as far as to invalidate agency action regarding land in which the public retains an interest, even where the agency action appeared to be specifically authorized by the legislature.[131]

Agencies should also be prepared to defend against actions by private parties in the nature of the common law writ of *mandamus* to compel them to take action to safeguard and manage public trust lands and resources. A mandamus action can only be maintained, however, where an agency is required to take action. If the agency has discretion either to act or not act, the challenge will usually fail.[132] A mandamus action compelling the attorney general to take action

---

127. See ch. 2, sec. H.

128. The situation would be different, of course, where an individual owns shorefront property in fee simple and the state attempts to restrict the owner's use under the public trust doctrine. The issue would be whether the trustee could impose a condition on the owner's use of his property related to the public's rights under the public trust doctrine. See ch. 2, sec. H.

129. For a discussion of delegation of public trust authority, see ch. 2, sec. D.

130. See, e.g. *Kootenai Environmental Alliance v. Panhandle Yacht Club Inc.,* 105 Idaho 622, 629–30, 671 P.2d 1085, 1092–93 (1983); *Meunch v. Public Service Comm'n,* 261 Wis. 492, 53 N.W.2d 512 (1952).

131. In *Gould v. Greylock Reservation Comm'n,* 350 Mass. 410, 215 N.E.2d 114 (1966), the court ordered the cancellation of a lease and management agreement between the state and private business to manage a ski area because the state agencies involved had exceeded their legal authority in consigning a public park to use as a large-scale commercial operation even though the Massachusetts legislature had expressly approved the development of a ski area. Although this case did not involve coastal areas, it is useful by analogy because of the characterization of public park land as "public trust" property.

132. See, e.g., *Lutheran Service Ass'n of New England, Inc. v. Metropolitan Dist. Comm'n,* 397 Mass. 341, 344, 491 N.E.2d 255, 258 (1986) (no mandamus action allowed to compel a park agency to take plaintiff's riverfront property, which the agency had originally intended to do in a comprehensive plan but later decided against).

to defend the public trust may also fail, because the attorney general has broad discretion to determine whether governmental intervention is appropriate.[133]

A state coastal agency may determine that it should take action to enforce public trust interests. The state as trustee has standing to bring a lawsuit to protect these interests.[134] Whether a particular agency or official can bring the action in the name of the state, however, must be determined based upon the scope of delegated authority in the agency's enabling statute and whether any statute authorizes the agency to sue and be sued.[135] Intervention by an agency may also be prevented if, by statute or case law, enforcement of public rights is entrusted exclusively to the state's attorney general. The state's attorney general is the traditional and typically best equipped and experienced enforcement arm of the state. Therefore, in any case, an agency contemplating enforcement of the public trust should consult and work with the state's attorney general.[136]

F. CONVEYANCE OF PROPERTY RIGHTS IN PUBLIC TRUST LANDS

This section addresses whether states can convey public trust lands to private owners and thereby extinguish the state's public trust interest. The section first considers judicial developments surrounding past grants by states to

---

133. See, e.g., *Secretary of Administration and Finance v. Attorney General*, 367 Mass. 154, 326 N.E.2d 334 (1975). If a plaintiff does convince a court to issue a writ of mandamus, the court will not instruct the public defendant as to how his discretion should be exercised, but only order him to exercise it conscientiously.

134. *State v. Deetz*, 66 Wis.2d 1, 224 N.W.2d 407 (1974). As noted in ch. 2, sec. C, a trustee has an obligation to take all steps prudently necessary to preserve the trust property and to enforce claims related to it.

135. See ch. 2, sec. D. An interesting example of state agency authority in this area appears in North Carolina General Statutes secs. 113–131(b) (3), which allows any state agency charged with protection of the public trust to initiate administrative proceedings to challenge decisions of other state agencies which violate public trust rights. In this connection, it should be noted that such a state agency should also be prepared to challenge, under applicable state law doctrines and under the federal consistency doctrine, decisions ignoring or adversely affecting public trust interests by federal agencies—including the Army Corps of Engineers. See generally ch. 4, sec. B, below.

136. An excellent model for any state that does not have clear or specific statutory authorization in this area is presented by North Carolina, which empowers the attorney general, upon request of the responsible state agency, to initiate a civil action for injunctive relief to restrain violations of the public trust and for mandatory orders to restore public trust resources to their undisturbed condition "[w]henever there exists reasonable cause to believe that any person or other legal entity has unlawfully encroached upon, usurped, or otherwise violated the public trust rights of the people of the state or legal rights of access to such public trust areas." N.C.G.S. secs. 113–31 (1987). It should also be noted that the attorney general in virtually every state has the common law power to institute actions to abate "public nuisances," such as the unlicensed obstruction of a navigable stream or other area encompassed by the public trust. In addition, state agencies charged with regulating development in public trust lands and waters are typically authorized to seek injunctive relief to restrain and remove unpermitted structures or fill. E.g., N.C.G.S. secs 43A–126(a).

private property owners for purposes of encouraging commerce (e.g., wharf-ing and filling statutes) and then examines the status of the law in several states with respect to the retained public rights in public trust lands that have been conveyed into private ownership. Finally, this section sets forth the standards derived from the principal judicial decisions governing the pro-cedures that states must follow in order to convey public trust land and to extinguish public trust interests.

May the state of its delegate agency convey lands held in public trust free of the trust obligations? From the seventeenth through the nineteenth centuries, many states conveyed tidelands to private parties to encourage commerce and naviga-tion. In some states literally hundreds of special statutes, often called "wharfing statutes," granted various interests in tidelands to private parties, generally au-thorizing them to construct and maintain wharves. As a practical matter, these statutes fostered extensive filling of tidelands for maritime and, later, urban development. During this period, legislatures apparently gave little thought to the state's trust obligations with respect to preserving public rights in the lands subject to the public trust, and they rarely included any explicit statement in these statutes to define the property rights being conveyed to the private parties or any rights being reserved for the public. Nor did any clear rules emerge from the courts with respect to these issues until the celebrated decision in *Illinois Central*,[137] which for the first time announced limits on a state's authority to convey public trust lands to private parties.

As noted above in chapter 1, section C, the Illinois legislature in 1869 had granted the Illinois Central Railroad full ownership[138] of 1,000 acres of the bed of Lake Michigan. This 1,000 acres comprised a substantial part of Chicago's down-town waterfront. The legislature afterward realized that it had gone too far and, in 1873, adopted an act revoking the grant to the railroad, which the railroad then challenged in a lawsuit. The United States Supreme Court held that the initial grant violated the public trust doctrine, was never valid, and therefore was revocable because the land was held in public trust and the state of Illinois had failed to carry out its duties properly in respect of the land. The Court stated that the public trust doctrine "requires the government to preserve such waters for the use of the public."[139] The court then established a rule that prohibits certain transfers of public trust land from the state to private parties:

> The control of the state for the purposes of the trust can never be lost, except as to such parcels as are used in promoting the interest of the public thereon, or can be disposed

137. 146 U.S. 387 (1892).

138. The state had granted a fee simple interest in the land, which is essentially the same as unqualified ownership.

139. Id. at 453.

of without any substantial impairment of the public interest in the lands and water remaining. . . . The state can no more abdicate its trust over property in which the whole people are interested, like navigable waters and soils under them, so as to leave them entirely under the use and control of private parties, except in the instance of parcels mentioned for the improvement of the navigation and use of the waters, or when parcels can be disposed of without impairment of the public interest in what remains, than it can abdicate its police powers in the administration of government and the preservation of the peace.[140]

The Supreme Court's ruling permits conveyance of ownership of public trust lands to private parties in limited circumstances. However, such a conveyance must serve a clear public interest and must not impair, or only minimally impair, the public's remaining interest in the shore. For example, a conveyance of public trust land that was filled long ago, has since been used for private purposes, and is physically no longer accessible to the shore might well be upheld. However, a conveyance of public trust land that still is capable of use for boating, fishing, recreation, and the like would probably not be upheld under the *Illinois Central* standard.[141] Under the Supreme Court's language, the possibility that an undeveloped section of the shore might one day be needed by the public suggests that a state could not grant unqualified ownership of such land to a private party. That is, any grant of such land would have to reserve the public's rights for future use of the land and resources.

Thus *Illinois Central* explicitly recognized the fiduciary duty of the states to protect tidelands and established a role for the judiciary to scrutinize closely any attempt by the state to terminate the public's rights. It did not, however, establish standards to guide or evaluate state dispositions of tidelands and left unclear such important matters as the nature of the public purposes that would justify conveyance to private parties, the threshold of impairment to the public's rights in remaining tidelands that would call a private grant into question, or the extent to which validly conveyed lands remain subject to public use under the public trust doctrine. Such omissions, and the relative lack of clarity on these issues in subsequent judicial decisions, obviously cause substantial uncertainty and concern among the real estate development, financial, and legal communities involved in coastal real estate activity.

In most states there is precedent to support the state's authority to extinguish the public trust absolutely, at least in certain circumstances.[142] The Supreme Court in *Illinois Central* was the first to suggest this, indicating that its decision

140. Id.

141. E.g., *Opinion of the Justices,* 383 Mass. 895, 424 N.E.2d 1092 (1981); *City of Berkeley,* 26 Cal.3d 515, 162 Cal. Rptr. 327, 606 P.2d 362.

142. Such authority was recently recognized by the United States Supreme Court in *Phillips Petroleum,* 484 U.S. 469, 475–76 (1988).

would depend, in part at least, on the explicitness of the legislature's statement of public purpose and the size of the parcel granted relative to the rest of the surrounding public trust area.[143] In 1926 the Supreme Court held that its reasoning in *Illinois Central* did not invalidate the city of New York's absolute grants of tideland and submerged land along the Hudson River to private owners.[144] The key difference in the later case appeared to be that the amount of land granted represented a small portion of New York's waterfront and that the city granted the land for a public purpose, the promotion of commerce.

The states have diverged recently over the possibility of absolute conveyances of public trust lands to private parties. A number of state supreme courts have held that absolute conveyances of former public trust lands to private parties are permissible: if the legislature expressly stated its intent to extinguish the public's interests; if the conveyance was made for a valid public purpose; and if the affected lands are no longer suitable for public trust uses (for example, because the area consists of landlocked tidelands filled in the nineteenth century and is isolated from the water).[145] The lower courts of Rhode Island have interpreted certain nineteenth-century Rhode Island Supreme Court decisions to hold that the filling of tidelands pursuant to governmental license and the legislative establishment of a harbor line extinguishes all public rights in the filled land.[146]

In contrast, the Vermont Supreme Court in 1989 all but eliminated the possibility of a legislative grant free of the public trust. In *Vermont v. Central Vermont Railway,*[147] the Vermont Court examined nineteenth-century legislative grants to littoral owners on Lake Champlain. Under these grants, the littoral owners had the right to erect wharves by filling submerged lands along the lakeshore to build railroad lines. The littoral owners, together with their heirs and assigns, were given the exclusive privilege of the use, benefit, and control of these wharves forever.

Under these legislative grants, the Central Vermont Railway's predecessor constructed a railroad line along a 1.1-mile strip of filled land centrally located on Burlington's waterfront. A century later, the Central Vermont Railway had only one active railway customer on the waterfront and contracted to sell much of the filled lands to a real estate developer. The court determined that although

143. 146 U.S. at 452–54.

144. *Appleby v. City of New York,* 271 U.S. 364, 393, 395 (1926).

145. See, e.g., *Opinion of the Justices,* 437 A.2d 597 (Me. 1981); *City of Berkeley,* 26 Cal.3d 515, 162 Cal. Rptr. 327, 606 P.2d 362 (1980); *Opinion of the Justices,* 383 Mass. 895, 424 N.E.2d 1092 (1981).

146. *DeLeo v. Nunes,* 546 A.2d 1344 (R.I. 1988), *app. den'd,* 109 S.Ct. 1522 (1989); *DeLeo v. Coastal Resources Management Council,* R.I. Superior Ct., C.A. No. 86-5127 (Aug. 9, 1988), *cert. den'd,* R.I. Supreme Ct., No. 88-388-M.P. (Dec. 9, 1988). The Rhode Island Supreme Court's reasoning, however, is inconsistent with *Illinois Central* and with the majority of state court decisions.

147. 571 A.2d 1128 (Vt. 1989).

Central Vermont Railway holds the title to the filled lands, its title is subject to a condition that the lands be used only for those purposes for which they were conveyed—railroad lines.[148]

A federal court also recently applied the criteria set forth in *Illinois Central* to invalidate a legislative grant of submerged lakebed involving a smaller tract of land,[149] despite the Illinois legislature's explicit finding that the grant would benefit the public. Loyola University of Chicago planned to fill 18.5 acres of submerged land to add to its campus, which would have provided 2.1 acres of public recreational area along the new shoreline and university athletic fields on the interior of the lakefill to which the public would have broad, but not totally unrestricted, access. The court first noted the presumption disfavoring attempts by the legislature to relinquish the state's power over public trust lands.[150] Second, Illinois law (like that of most states) narrowly construes any conveyance of public trust lands to a private entity.[151] Finally, any legislative grant of trust land violates the public trust doctrine if the grant's primary purpose is to benefit a private interest.[152]

Loyola argued that the lakefill would increase the public's access because the public had no access to the existing privately owned lakefront area. The court rejected this argument, noting the public had unrestricted access to the submerged lands at issue, which lands would be reduced by the filling.[153] The court emphasized the grant's private goal of enlarging the Loyola campus, holding that public benefits of this grant, no matter how great, would be merely incidental to Loyola's benefit.[154]

Although the standards applicable from state to state are neither uniform nor absolute, the following six criteria to determine the propriety of a conveyance public trust lands by a legislature have emerged:[155]

---

148. Id.

149. *Lake Michigan Federation v. United States Army Corps of Engineers,* 742 F.Supp. 441 (1990).

150. Id.

151. The court noted that Illinois courts have upheld only one grant of trust lands to a private entity of trust lands. In that instance, the legislature conveyed several acres of submerged land to a local parks commission in order to construct an extension of Lake Shore Drive. The legislature authorized the parks commission to fill and convey land in between the roadway extension and the existing shoreline to upland owners in order to defray the costs of building the extension. *People v. Kirk,* 162 Ill. 138 (1896).

152. Id.

153. Id.

154. Id.

155. These criteria are a consolidation of factors discussed in *Opinion of the Justices,* 383 Mass. 895, 424 N.E.2d 1092, and *City of Berkeley,* 26 Cal.3d 515, 162 Cal. Rptr. 327, 606 P.2d 362. See also *Scott v. Chicago Park District,* 66 Ill.2d 65, 360 N.E.2d 773 (1977); *New York Power & Light Corp. v. State,* 245 N.Y.S. 44 (1930); *Obrecht v. National Gypsum Co.,* 361 Mich. 399, 105 N.W.2d 143 (1960); *Thomas v. Sanders,* 65 Ohio App. 2d 5, 413 N.E.2d 1224 (1979).

1. The grant must refer explicitly to the land in question. The legislation must describe the land granted with particularity, similar to the description found in a deed or mortgage.
2. The legislature must acknowledge the public interests being surrendered. A grant of submerged land—for example, for wharfing out or for dockominiums—will not affect the right of the public to use the land for fishing, navigation, or other purposes consistent with public trust rights, unless the legislation expressly extinguishes the trust.
3. The grant must recognize and identify future uses. Thus grants of public trust land which have not expressly provided that the private owner or his successors use the land for any particular purpose have been construed to be subject to a condition that private owners must use the land for a public purpose.
4. The conveyance must serve a valid and articulated public purpose. Environmentalists and private property owners will undoubtedly disagree over what constitutes a valid public purpose, and few generally applicable guidelines have been developed by the courts thus far. Different courts purportedly applying public trust principles have decided, for example, that use of a section of New York City waterfront for commercial development is an acceptable public purpose,[156] but development of a commercial resort on undeveloped land is not.[157]
5. The grant of public trust land will probably be subject to less scrutiny when the land in question is in fact no longer suitable for public trust uses. For example, courts may be more likely to uphold extinguishing public trust rights in tidelands which were filled decades or centuries ago and which have been built upon and no longer abut the coast.[158]
6. In accordance with the essential holding in *Illinois Central,* the conveyance must not cause or contribute to a substantial impairment of the remaining public trust lands or of the public's rights and interests in them.

G. LICENSING AND LEASING

Once states have identified their public trust areas, they may consider the scope of their authority to grant licenses or leases to private entities to use them. Licensing and leasing can serve both a regulatory and a revenue-generating function. This section describes the practical considerations to

156. *Appleby,* 271 U.S. at 391.
157. See *Greylock Reservation,* 350 Mass. 410, 215 N.E.2d 114.
158. See *Opinion of the Justices,* 383 Mass. 927, 424 N.E.2d 1111 (1982).

be taken into account in establishing a licensing or leasing program for public trust lands, waters, and resources and the basic requirements for such programs.

One of the most important questions in public trust law concerns the extent of the relevant agency's power to issue licenses or leases for activities on public trust lands.

The grant or denial of licenses (or permits)[159] is perhaps the most common form of discretionary administrative action and accounts for the bulk of the administrative process generally. This is as true with regard to the activities of coastal managers, including activities related to protection of the public trust in coastal resources, as it is of other agencies. Licensing is also potentially the most drastic of an agency's procedures. Other agency actions, such as rate setting, imposition of penalties, fines or orders of reparation to third parties, the grant or denial of a government bounty, or the distribution or withholding of an insurance fund, can be absorbed by those affected as a business cost or otherwise adapted to. Absent a required license, however, the activity that must be licensed cannot be started or continued at all.

All states have long recognized the authority of the state legislature to enact statutes providing for the granting or denial by specified agencies of licenses as part of the state's police power to regulate private activities in the interest of public health, safety, and welfare. Licensing may be used for three different regulatory purposes, all of which are relevant to coastal management: to limit entrance into or competition in a field or activity (e.g., an exclusive license or concession or a monopoly) on the theory that the public is thereby best served; to establish the conditions and standards of performance for the license recipient (the licensee); and to police the licensee's performance. Most agency licensing is a combination of rule-making and adjudicatory processes involving (through the application to particular fact situations of standards established either by the legislature or by the agency through regulations) a determination of whether an applied-for activity is in the public interest and whether the particular applicant is eligible to receive the license sought. The process frequently also involves the subsequent determination of whether the licensee is continuing to meet applicable standards, requirements, and conditions so as to remain eligible to enjoy the benefits of the license or whether its failure to do so requires suspension or revocation of the license.

Not only has licensing been a distinctive feature of administrative agency

159. There is no significant legal or functional distinction between licensing and permitting, despite different usages from state to state. Both essentially involve the legal permission to engage in conduct or activity that would otherwise be improper or illegal. For convenience, and because most state administrative procedure laws use the term *license,* that term is used here.

activity generally in this country, but the licensing of activities affecting tidelands was in fact the prime motive for the creation of one of the earliest state administrative agencies.[160] The control of coastal and waterfront development through licensing programs has continued to be a distinctive feature of state regulation of business activity. This regulatory activity has intensified in recent years as the result of popular concern and agitation over the potentially harmful consequences of unregulated or uncontrolled private activities to the environment and to scarce natural resources.

A number of issues relevant to coastal management licensing programs deserve specific discussion.

1. Initial Licenses

Traditionally, at common law in virtually every state, the granting of a license was deemed to be the granting of a mere privilege, and no one was regarded as having any right or property interest in obtaining or retaining a license. Consequently, unless a statute required otherwise, the state (through a properly delegated agency) was free to grant, deny, suspend, or revoke a license in any way and for any reason it wished, without any right to notice or a hearing.

However, this is no longer the case. Recent judicial recognition of the constitutional right of a licensee to procedural due process before the license can be taken away, as well as explicit provisions in most state administrative procedure acts, have resulted in greater protection for the rights of existing licensees in the form of requiring a full-fledged adjudicatory hearing. (See below, subsection 2). However, virtually no statutes prescribe any minimum procedures to be followed by licensing agencies in the issuance or denial of a license in the first instance. In the absence of any statutory requirements, agencies continue generally to be free to establish initial license processes with as many or as few procedural safeguards or rights for the applicant as they deem appropriate. Consistent with this situation and with the basic theory of the public trust—that public trust lands are public property and a license to use or develop public trust lands confers only a conditional privilege on a private person to use public property—there is no right to an adjudicatory hearing should a license application be denied.[161]

160. For example, in 1866 the Massachusetts legislature, concerned about the effects of unfettered wharfing out and filling of tidelands and other private waterfront development, created a permanent Board of Harbor Commissioners to regulate development in Boston Harbor by allowing the conduct of waterfront activities only pursuant to tidelands licenses. 1866 Mass. Acts c. 149.

161. It is interesting to note that, for greater protection of the public interest and to further public participation in the regulatory process, some statutes and regulations provide for any aggrieved citizen or other person to request an adjudicatory hearing on, or to appeal, a decision by an agency to *grant* such a license. E.g., Mass. Gen. L. ch. 91, sec. 18. It should also be noted that there are certain kinds of technically initial license applications which reflect sufficiently protectable interests so as to be entitled to the procedural safeguards discussed in subsection 2, below, such as the holder of a valid

2. License Revocation or Suspension

As noted above, as a result of constitutional requirements of procedural due process and specific statutory requirements in most states, an existing license cannot be revoked, suspended, or refused renewal without notice and an opportunity to be heard in a trial-type (adjudicatory) proceeding.[162] The specific procedural requirements mandated by statute or constitution may vary from state to state, but any coastal agency should be safe from reversal for procedural errors if it adopts and applies the following rules for hearings on license revocations, suspensions, or renewals:

All parties will be afforded the opportunity for a full and fair hearing prior to the agency's decision.

All parties will be given sufficient advance notice of the particular issues involved in the hearing.

All parties will be accorded sufficient notice of the time and place of the hearing.

Although courtroom rules of evidence need not be observed, rules of evidentiary privilege will be honored, and all relevant evidence that is not unnecessarily repetitious or duplicative will be admitted.

All parties will have the right to call witnesses, examine and cross-examine them, and introduce written and rebuttal evidence.

All parties will have the right to have the agency subpoena relevant witnesses and documents.

All evidence, including records in the agency's possession, which the agency will use in reaching its decision must be offered and made part of the record.

The agency will not take "judicial notice" of matters within its special knowledge unless it notifies the parties of the matters so noticed and gives them an opportunity to contest the facts so noticed.

The agency will not reach conclusions or make findings of fact based upon its expertise or special knowledge unless they are supported by some evidence in the record supplied by expert witnesses.

The agency will prepare an official record, including testimony and exhibits (but this record may be in narrative form, and the agency need not transcribe testimony unless requested by a party, who must pay the cost of the transcription).

---

license for a particular use or structure, or a grandfathered noncomplying preexisting occupant, who proposes a change in the current use or structural alterations and who must, under applicable regulations, apply for a new license; or the owner of an unlicensed or improperly licensed structure who must comply with a new licensing scheme to continue use.

162. Typically, statutes and the requirements of due process exempt from the mandate of a prerevocation hearing situations where the agency is required by law to revoke a license without exercising any discretion in the matter, as for example on the basis of the licensee's conviction of a felony and situations involving a licensee's unjustified failure to submit timely reports, pay prescribed fees, or maintain insurance coverage as required by any law or regulation.

Where a majority of agency officials who are to make the final decision have not heard the evidence, any recommended or proposed decision by a duly delegated subordinate official after a full adjudicatory hearing will be presented to the parties, who will be given a reasonable opportunity to file objections and present arguments against it.

Each agency decision will be in writing and accompanied by a statement of reasons for the decision, including the determination of each issue of fact and law necessary to the decision.

Along with notification of the agency decision, parties will be notified of their rights to administrative review or judicial appeal and the time limits involved.

A central feature of judicial review of agency decisions revoking licenses is the requirement imposed by most statutes that key agency findings of fact, or the agency licensing decision generally, be supported by "substantial evidence" in order to be upheld by the reviewing court. "Substantial evidence" is often unhelpfully defined as such evidence as a reasonable mind might accept as adequate to support a conclusion. However, the standard is not really susceptible of either simple definition or easy application. Certain things are nonetheless clear. The substantiality of evidence must be based on all of the evidence in the administrative record, not just that which supports the agency's conclusions. The agency's decision must accordingly address and take into account contradictory or adverse evidence or any other evidence in the record from which conflicting or contrary inferences could be drawn.

However, the conclusions of the agency officials conducting the hearing as to the credibility of the witnesses who appear before them are entitled to substantial deference from reviewing courts, and the agency's choice between conflicting testimony or whether to believe even an uncontradicted witness will not be second-guessed by a reviewing court, even if the court would independently have reached a different result. Similarly, the agency is entitled to draw reasonable inferences from the evidence before it, which are entitled to deference from the courts when based upon the specialized knowledge and experience of the agency officials. Finally, the function of the reviewing court is not to exercise an independent assessment of the evidence or substitute its view of the evidence for that of the agency, even if different factual conclusions might more reasonably or logically have been reached through evaluation of the same evidence and the use of other reasonably drawn inferences, so long as there is substantial evidence in the entire record to support the agency's findings.[163]

163. Since it is a recognized principle of administrative law that an agency may adopt general policies through either rule making or case-by-case adjudication [e.g., *SEC v. Chenery Corp.*, 332 U.S. 194 (1947)], it may be preferable in particular cases for an agency to establish policy by regulation

3. Leasing

A significant amount of coastal agency business with respect to authorizing tidelands uses and activities is currently done in many states in the form of leases rather than in the form of licenses.[164] Because a lease is essentially a legal authorization by the owner of land for the possession and a particular use of the real property by a lessee for a certain fixed period of time, it is really not functionally different from a license permitting such use, with the principal difference being the fact that a lease is typically renewable or terminable in accordance with its provisions rather than by the provisions of statutes requiring procedural due process for licenses (which may argue for seeking licensee rather than lessee status on the part of those seriously concerned about potentially arbitrary conduct by an agency disrupting legitimate economic expectations).[165]

Even where an agency's enabling act does not expressly authorize it to grant leases, if it has general licensing authority with respect to activities subject to its jurisdiction, the power to lease as well as license or in lieu of licensing will almost certainly be implied (see ch. 2, sec. D). Leases by public trust agencies in coastal areas should accordingly be upheld as long as they are granted pursuant to established regulatory criteria, are consistent with the purpose of the regulatory scheme administered by the agency, are not in conflict with any statutory restrictions (as, for example, express limitations on the length of licenses), and are adequately protective of the public's proprietary interests (e.g., contain proper

---

because of the more lenient judicial standard for review of rule making, i.e., the regulation will be upheld as a valid exercise of discretion under a standard of minimum rationality, as long as it is not entirely wrong and is supported by some significant evidence. See ch. 2, sec. D, discussing delegation and rule-making authority. Whether the coastal agency proceeds by regulation or by adjudication, however, it should seize every opportunity to include in its regulations and decisions, regarding licenses generally or in particular instances, an explicit finding that its action is consistent with and in furtherance of public trust principles which are embodied in its enabling act or in its regulations (if the enabling act is vague or silent) which it is obligated to implement by virtue of its delegated responsibilities. See ch. 3.

164. Leasing has, of course, long been a feature of state regulation of submerged lands (lying seaward of the mean low watermark), particularly since the Submerged Lands Act of 1953 [43 U.S.C. secs. 1301–56 (Amend. 1986)], with respect to exploration and drilling for oil, natural gas, and other mineral resources.

165. The basic distinctions between leases and licenses in most states are legal niceties of little relevance to coastal managers, in that a lease is technically the conveyance of an interest in land and must be in writing (and sometimes under seal) to comply with ancient requirements (particularly the Statute of Frauds). Traditionally, the essential characteristic of a license that distinguished it from a lease was that it was revocable at the will of the owner or rightful possessor of the land subject to the license, a distinction that is no longer meaningful in light of statutory and constitutional safeguards currently accorded licensees (see above, subsection 2). Since many, if not most, coastal licenses are for substantial terms of years, in order to justify licensees' frequently large capital expenses and to make financing for construction or improvements feasible, they have come to be equivalent to leases, and the distinction between a modern coastal licensee and a lessee is of little practical significance.

covenants to guard against waste or other misuse). In these respects they do not differ in any meaningful way from equivalent licenses as a coastal management technique.[166]

Whether a coastal agency's grant of permission to a private party to use, build, or conduct an activity on lands subject to the public trust takes the nominal form of a license or a lease, perhaps the key element of the agency decision (besides insuring that the activity is consistent with public trust purposes and is subject to procedural safeguards for the public's interest) is the length of the governmental permission. Although much nineteenth-century tidelands licensing activity was regarded by most licensees as effectively conferring rights in perpetuity (or the equivalent of fee simple title), virtually all states currently impose some sort of time limit on coastal leases or licenses, ranging from terms of a year or two to fifty or sixty years or more, and require some sort of meaningful renewal procedure. Statutory limitations of terms, to the extent they exist, must of course be strictly observed and implemented by the coastal agency. In the absence of such restrictions, the agency has considerable discretion regarding the setting of lease or license terms. Several principles are available to guide the agency's discretion in this regard.

First, since the object of such leases and licenses is of great public importance and subject to potential controversy among conflicting interests, the prudent agency should definitively establish the terms of years for the various categories of coastal projects and uses (as well as the conditions pertaining thereto) by promulgation of regulations. There is a certain degree of judicial preference for seeing rules or standards of general application created by regulation rather than ad hoc adjudication, despite agency discretion to proceed either way, since rule making forces important issues to full public display and in that sense makes for more responsible administrative action. In addition, as noted in subsection 2 above, from the agency's perspective, its rules will be subjected to less rigorous judicial scrutiny than agency adjudicatory decisions and are consequently more likely to survive intact if challenged.

Second, the agency should presume that the granting of rights to use or build on lands subject to the public trust should be for a fixed and limited period of time, subject to renewal, and not in perpetuity (although leases or licenses for certain noncommercial, water-dependent uses either of general public benefit, such as a wholly public wharf, or of great expense and permanence, such as a harbor tunnel for traffic or utilities lines, or of relative permanence and minor

166. Under normal principles of private trust law, a trustee of real property has an implied power to make leases, even if the document creating the trust did not specifically so provide, as a means of both protecting the property and/or making it productive, as long as the terms of the leases are reasonable under all the circumstances and do not prejudice the rights of beneficiaries. See *Restatement (Second) of Trusts* 189 (1959).

public impact, such as a small dock for use by a preexisting private residence, might be optionally allowed for unlimited terms, subject always to the agency's right to modify in light of the public interest or for failure to comply with reasonable conditions). This principle reflects the fact that an expiration procedure should be intrinsic to the concept of a coastal lease or license. The coastal agency, in discharging its public trust responsibilities, must have the opportunity to reevaluate the public trust lands uses and structures in light of changing circumstances and standards of proper public purposes. Moreover, the agency is itself obligated to ensure that lessees and licensees are operating in accordance with statutory and regulatory requirements and with all original lease or license conditions, as well as with proper recognition of the importance of the shoreline and its related resources as irreplaceable assets of benefit to the entire public.[167]

In addition, however, the availability of a renewal procedure upon expiration of a fixed term is also essential to responsible coastal leasing and licensing. Lessees and licensees who must often expend large sums to build or maintain coastal facilities would have little incentive in most cases to do so without the prospect of the ability to renew, and might even be unable to undertake their activities in the first instance because of the unavailability of funds for projects whose life spans are too short to repay lenders. Consequently, coastal agencies must be cognizant of the realities of obtaining financing and legitimate investment expectations in establishing lengths of leases or licenses for particular uses. No rules of general application can be laid down, beyond a rule of thumb that the larger, more complicated, and more expensive the cost of building and maintaining a tidelands project for a proper public trust purpose, then the longer, presumptively, should be the lease or license term (sometimes possibly requiring a term of as long as ninety-nine years to be commercially feasible). The burden of persuasion nonetheless should be placed on the proponents of long terms to justify their practical necessity, and the agency must be constantly alert to avoid the preemption of projects requiring long terms for maximum public benefit or of feasibly concurrent uses by activities that need to be allowed only on a short-term basis.[168]

167. The failure of lessees or licensees to maintain substantial compliance with reasonable public interest conditions imposed by the coastal agency or to operate so as to substantially advance or protect the public purposes on the basis of which the lease or license was originally granted will justify the agency in revoking the lease or license. E.g., *Smith v. New York*, 545 N.Y.S.2d 203 (App. Div. 2d Dept., 1989); *Boston Waterfront Development Corp. v. Commonwealth*, 378 Mass. 629 (1979). Such revocation, or permit rescission, will not constitute a compensable taking of private property when justified by public trust considerations. E.g., *Marine One, Inc. v. Manatee County*, 898 F.2d 1490 (11th Cir. 1990).

168. It should be recognized that, if a statute is silent as to an agency's authority to limit the terms of leases or licenses, it is inevitable that someone will argue that the agency has no such authority. The general doctrines of agency powers (see ch. 2, sec. D) and the public trust doctrine should be

4. Fees

It is clear that a state legislature may properly delegate to an agency charged with administering a licensing or leasing program the power to establish and impose through rule or regulation a system of license fees as a valid exercise of the state's police power to regulate particular businesses or activities in the public interest, and a great many states in fact have done so in connection with coastal regulatory programs. Where specific fee schedules are not mandated by the legislature, the general rule as to "regulatory" fees pursuant to the police power is that the amount of the fee should be commensurate with the reasonable expenses incident to administering the licensing program. Some courts may be relatively strict in confining the amount of charged and collected fees to the recovery of the actual costs incurred by the agency in conducting its licensing/leasing activities; others may be much more liberal in interpreting the extent to which the fees must match actual costs and may be disinclined to scrutinize the fee structure too closely as long as it is within a range of reasonableness and as long as any incidental surplus revenue is used for a proper agency purpose.

The basic distinction to be aware of in this area is between the power to impose fees and the power to tax. In virtually every jurisdiction the power to tax, typically arising out of distinct and explicit language of state constitutions, is held to be a uniquely legislative power that is either totally nondelegable or delegable only in very limited circumstances. The courts rarely allow the police power, however broadly it is defined, to become so distended as to include or be a substitute for taxation. Consequently, administrative agencies presumptively may not impose taxes in the guise of license fees. A fee is essentially distinguished from a tax in several ways. It is charged in exchange for a particular governmental service that benefits the person paying the fee in a manner not shared by other members of society. It is paid voluntarily, in the sense that the person paying the fee has the option of not using the service and thereby avoiding the charge. And most significantly, it is imposed not for the purpose of raising revenues but rather to compensate the governmental agency providing the service for its expenses.[169]

Coastal managers have more flexibility than most administrative agencies in the area of structuring licensing/leasing fees, because they are not merely regulating activities under the police power but are also regulating the use of lands

---

sufficient to rebut such an argument. Similarly, an argument is certain to be made that imposing fixed terms on existing uses may amount to a taking of property without just compensation. This position should not succeed either, given the proprietary basis for the public trust doctrine. See generally the discussion of the law of takings in the public trust context, in ch. 2, sec. H.

169. See, e.g., *Emerson College v. Boston,* 391 Mass. 415 (1984).

subject to the public trust, i.e., public property. They therefore can impose "user" fees in addition to "regulatory" fees and establish fee schedules with the legitimate intention (and sometimes legislatively mandated obligation) of deriving full and adequate financial return from those seeking to occupy, use, or obtain easements over public lands, above and beyond the costs of administering the regulatory program, in order to derive just compensation for their activities on tidal, former tidal, and submerged lands.[170]

Coastal managers, within whatever specific limitations have been set by the legislature, have wide latitude in establishing fee structures to secure such compensation, including, as examples: increasing prior fee levels for tidelands occupation that are marginal or inadequate by present standards; charging full, fair market rental values for the term of a lease or license as established through an appraisal process; imposing periodic adjustments to reflect the impact of inflation; setting reduced rates or providing grace periods or "grandfathering" for existing unauthorized water-dependent uses in order to give them a reasonable opportunity to come into compliance with new, more rigorous regulations; exempting municipalities and other governmental activities; and adjusting fees for particular activities so as, for example, to encourage water-dependent uses, traditional waterfront enterprises, or projects that enhance public access or benefit, while discouraging proposals for the filling or dredging of flowed tidelands or non-water-dependent uses. Similarly, managers have considerable discretion as to the methods by which fees are assessed, as for example choosing to impose lump sum fees rather than annual fees, or a combination of both, or to require appraisals in particular instances.

Notwithstanding substantial discretion in the area of license and lease fees, coastal agencies must, in the development and implementation of fee schemes as in all their other administrative responsibilities, remain ever cognizant of the peculiar difficulties they face in dealing with and choosing among numerous valid yet often competing public trust activities.[171] Discretion must be exercised

170. Whether the revenues thus obtained can be retained by the agency for use in its general administrative activities, or must be earmarked for some special purpose (e.g., support of public education), or possibly even returned to the state treasury for general purposes is a matter that will vary from state to state depending upon specific legislative direction. It should be noted that the proprietary, as well as police power, nature of a coastal agency's regulatory power with respect to public trust lands is also sufficient to justify the agency's imposition of virtually any condition upon a lease or license for activities affecting such lands—including such common, commercially reasonable conditions as obtaining performance bonds or letters of credit to assure project completion or establishing escrow accounts or financial reserves to assure ongoing project compliance—as long as the condition reasonably advances or protects public trust interests and public purposes, including the maximization of public access to and enjoyment of beaches and waterfronts and of water-dependent uses.

171. See generally the discussion of public trust uses in ch. 2, sec. B for a more thorough analysis of this ever-present problem.

with sound judgment that is explicitly sensitive to the perplexing though necessary task of balancing legitimate but conflicting interests in the tidelands. For example, an agency may justifiably seek to set fees for new marinas at relatively high levels to reflect the unique value of the benefits such uses derive from exclusive, long-term occupation of the public lands; but it must also be careful not to set them so high that marinas go out of business or recreational boaters are driven to out-of-state facilities, thus arguably defeating the goal of public access.[172]

5. License Coordination

Few, if any, coastal agencies operate in splendid isolation or draw on clean slates in implementing their management programs. The recent flood of environmental and development control legislation in almost every state has created a confusing array of multiple and often overlapping license requirements, particularly in long-settled urban waterfront areas. From the perspective of private developers and other would-be licensees, a license "crisis" appears to confront them, with the burden and expense of obtaining so many local, state, and federal permissions sometimes so overwhelming that they give up in frustration. Problems of multiple regulatory requirements reflect the characteristic fragmentation and decentralization of American governmental authority and are present in all areas of land use planning and development, with coastal management no exception.

A coastal agency may be clearly delegated public trust responsibilities generally but nonetheless may have to coexist with another separate agency that, for example, must review all projects which significantly impact the environment; or one that is empowered to control mineral extraction; or one with authority over all public utility transmission lines even in coastal areas; or one specifically charged with approving the construction of nuclear energy facilities (which frequently need access to large bodies of water for cooling purposes); or one that regulates the disposal or removal of toxic or hazardous wastes which have often resulted from past uses of waterfronts; or one with authority over shellfish gathering in tidal areas; or one responsible for regulating scientific research in tidal or submerged areas; or one with authority to preserve the historic and archaeological resources that are frequently found along our coasts and waterfronts; or with several or all of such other agencies. Moreover, the coastal agency may also have to deal with another state agency that shares public trust responsibilities,[173] as

---

172. A similar dilemma regarding efforts to protect and repair beaches by imposing access or parking fees that are so high that their use is discouraged is discussed in ch. 3, sec. C, subsection 1.

173. See the discussion of delegation in ch. 2, sec. D. It should be noted that some coastal managers are of the opinion that courts tend to be more favorable to agencies charged with regulating the state's own lands than to agencies charged with regulating activities on the land of others,

well as inevitably having to function in conjunction with federal agencies, such as the Army Corps of Engineers and the Environmental Protection Agency, that regulate various aspects of tidelands and coastal wetlands activities.[174]

This multiplicity of regulatory jurisdiction in the coastal area obviously underscores the need for interagency cooperation and coordination of licensing procedures, in order to minimize duplication of effort and expense as well as to expedite the processing of project applications.[175] Some states have attempted to address the issue structurally by empowering a particular state official or agency to resolve jurisdictional or procedural disputes between agencies and to harmonize conflicting agency regulations at the behest of an agency or an affected applicant.[176] Various other states have created a state coordinating agency or official charged with orchestrating agency and applicant participation in a coordinated licensing process. Under such statutes, coordination may be either required or optional; may or may not involve adjudicatory hearings; may or may not take the form of joint hearings on all the licenses required for a particular project; may or may not authorize the coordinating agency to develop a master application form to be used and completed by all of the relevant agencies; may or may not provide for waiver of jurisdiction by any agency that fails to respond to the master application within a specified time period; and may or may not empower the coordinating agency to issue a single "one-stop" license that satisfies all or a specified number of statutory license requirements applicable to a project. What they all have in common is a commitment to expediting the license process to reduce time and review costs.[177]

Most states do not, however, have statutes expressly dealing with the coordination problem in coastal project licensing (or other areas of land use regulation and control). In such circumstances, assuming the infeasibility of successful

---

leading to the desirability of careful strategic coordination when seeking to vindicate the public trust in novel or controversial situations.

174. It is arguable and probable that, to whatever extent an agency has regulatory authority in or over lands subject to the public trust, the agency must accordingly discharge its regulatory responsibilities not only in accordance with public trust principles, but also with the recognition of its obligation to make decisions affecting such lands only after taking those principles adequately into account.

175. The problems of federal-state coordination are, of course, specifically addressed in the national Coastal Zone Management Act, particularly its consistency provisions. The act has, since 1980, declared license simplification and coordination to be a national policy objective. 16 U.S.C. sec. 1456(c), (h) (1972). See generally ch. 4.

176. E.g., Mass. Gen. L. ch. 30, sec. 5 (1919).

177. Washington was the first state to adopt license coordination legislation in 1978. Wash. Rev. Code Ann. sec. 90.62. See, e.g., Or. Rev. Stat. secs. 447.800–447.865; Alaska Stat. secs. 46.35.010–46.35.210; Fla. Stat. Ann. secs. 288.501–288.518; Minn. Stat. Ann. secs. 116C.22–116C.34; Md. Code Ann. art. 78A, secs. 56–67, for various examples of coordination statutes.

sponsorship of such legislation, coastal agencies with the necessary desire are not without means to address this issue themselves. As long as they are acting within the broad scope of their individual statutory authorities and in furtherance of their particular delegated purposes, there is no reason why agencies with shared or overlapping responsibilities for lands subject to the public trust and common interests in saving public moneys (and their own resources) by simplifying license requirements and processes cannot agree to work together, either informally or by formal regulation, to facilitate the coordination of coastal licensing. Efforts at such self-initiated coordination have recently been undertaken in several states by means of joint review panels or interagency task forces composed of relevant local, state, and/or federal agencies banding together to review such matters as offshore mineral or other development proposals with a view to minimizing conflict among competing legitimate uses (e.g., fishing vs. oil drilling), as well as reducing time and expense by coordinating the license application review processes.[178] At the very least, such agencies can cooperate in creating or supporting mechanisms such as the permit advisory services established by a number of state coastal zone management agencies to provide license applicants with sufficient information (such as necessary forms, explanatory brochures describing the regulatory regimes different types of coastal projects or uses must comply with, advice as to filing and timing, availability of opportunities to obtain staff guidance, etc.) so that the applicants themselves can contribute to the streamlining of the process.

Such interagency or intergovernmental coordination programs, even on a limited basis, are not easy to devise or implement successfully. In many instances, agencies will be jealous of their statutory prerogatives and resist what they perceive to be invasion or surrender of their jurisdiction. Local histories of interagency competitiveness and "turf wars" may complicate the effort. Many procedural difficulties may be encountered, such as coordinating the license process of agencies that require formal adjudicatory hearings with those that do not. Particularly difficult to resolve will be the almost inevitable rivalry between agencies that emphasize environmental and conservationist priorities in the discharge of their public trust responsibilities and agencies that are more receptive to developmental public trust uses, with each group fearing that its special concerns will not receive sufficient consideration in a coordinated proceeding. Despite all such problems, however, until statutory rationalization creates a public

178. See Hildreth, "The Public Trust Doctrine and Conflict Resolution in Coastal Waters: West Coast Developments," *Proceedings of the Sixth Symposium on Coastal and Ocean Management,* pp. 2613–2615 (1989).

trust "czar" or superagency, license coordination and simplification should be a goal for consideration by every coastal agency, if only in its self-interested pursuit of efficiency.[179]

H. INTERACTION OF THE PUBLIC TRUST DOCTRINE WITH THE TAKINGS CLAUSE OF THE UNITED STATES CONSTITUTION

The public trust doctrine most often confronts private property owners who claim rights in public trust lands. Because of these claims, actions and regulations under the public trust doctrine may be challenged by private landowners as takings without just compensation in violation of the United States Constitution and state constitutions. This section briefly reviews the development of takings law and the United States Supreme Court's modern takings decisions, including *Lucas v. South Carolina Coastal Council*.[180] It argues that a state's assertion or reassertion of its public trust authority under the public trust doctrine in order to regulate or control activities in public trust areas, because it is property-based authority, will successfully withstand challenge as an unconstitutional taking. This section also considers the issue of state regulation of public trust lands, distinguishing between regulation that is consistent with the public trust doctrine and that which is more restrictive than the doctrine requires. Finally, this section addresses the question of takings challenges to state actions either to expand the geographic scope of the public trust doctrine beyond its traditional boundaries or to provide access to public trust lands over private property.

1. Background of the Takings Doctrine

The Fifth Amendment of the United States Constitution provides, in part: "No person shall . . . be deprived of life, liberty or property, without due process of

179. Most difficult of all coordination situations will be those that necessarily require regional, multistate cooperation for the effective achievement of their coastal management objectives, such as the massive effort to protect and plan for the future of the Chesapeake Bay, which directly involves three states and the District of Columbia and all of their relevant agencies, as well as the federal government and its numerous agencies that regulate or occupy land in the area, and also must eventually implicate at least three other inland states containing large portions of the bay watershed. The sheer complexity and duration of such programs may make informal, cooperative approaches insufficient and require parallel statutory procedures agreed to by all of the jurisdictions involved and, most probably, an interstate compact approved by Congress to ensure ongoing and effective planning, permitting, and coordination.

180. 112 S.Ct. 2886; 120 L.Ed.2d 798.

law; nor shall private property be taken for public use, without just compensation." The Fifth Amendment applies to the various states through the Fourteenth Amendment, which provides, in part: "No state shall make or enforce any law which shall abridge the privileges or immunities of citizens of the United States; nor shall any state deprive any person of life, liberty, or property, without due process of law; nor deny to any person within its jurisdiction the equal protection of the laws." The Fifth and Fourteenth Amendments were traditionally applied only in the case of physical occupation of private property by government entities (i.e., when the government occupied the property and dispossessed its owner). However, beginning in the early 1900s, courts began to recognize takings claims when government so severely regulated or restricted the use of private property that the owner was effectively deprived of any use or economic value of the property, tantamount to physical occupation. Such restrictions were deemed "regulatory takings."

The United States Supreme Court has repeatedly stated that takings analysis is essentially an ad hoc enquiry into the facts of each case. Therefore, understanding the takings doctrine requires a review of a series of major Supreme Court takings decisions during the past seventy years defining the parameters of the doctrine as applied in specific factual circumstances.

The Court laid the groundwork for the modern doctrine of regulatory takings in 1922 in a challenge to Pennsylvania's Kohler Act, which required coal companies to leave deposits of coal in sufficient quantity in order to prevent subsidence of land and damage to property.[181] The Supreme Court found the Kohler Act was an unconstitutional taking without just compensation because it made the mining of certain coal deposits "commercially impracticable." In what has become a famous and frequently quoted holding, the Court stated "[t]he general rule at least is that while property may be regulated to a certain extent, if regulation goes too far it will be recognized as a taking."[182] The Court did not, however, define what it meant by going "too far."

In 1978 the Supreme Court reexamined its approach to takings claims and imposed a degree of order in takings analysis. The New York City Landmarks Commission's denial of a plan to build a fifty-story office tower above Grand Central Station was challenged as a regulatory taking.[183] Finding that no taking had occurred, the Court stated that regulatory takings claims require "essentially ad hoc, factual inquiries." The Supreme Court established a framework for takings analysis consisting of three elements to guide these inquiries. In deciding

181. *Pennsylvania Coal Co. v. Mahon,* 260 U.S. 393, 415 (1922).
182. Id. at 415.
183. *Penn Central Transportation Co. v. City of New York,* 438 U.S. 104 (1978).

takings claims, courts must consider: the economic impact of the regulation upon use of the property; the extent to which the regulation interferes with the property owner's "distinct investment-backed expectations"; and the character of the governmental action (whether it allows a "physical invasion" of the property by the public or merely adjusts the burdens or conditions of our national economic life among broad classes of citizens).[184]

In 1987 the Supreme Court reviewed three regulatory takings claims.[185] In the first case, applying the framework takings analysis it had established in *Penn Central,* the Supreme Court found that a Pennsylvania subsidence statute requiring that a small fraction of the total amount of coal available for mining be left in the ground to prevent surface subsidence did not constitute a taking.[186] In the second case, the Court declined to decide the taking claim, but held that a "temporary" taking (from the date a regulation restricted use of property to the date of decision) could entitle a property owner to compensation.[187] In the third, the Court found that the state's requirement that a private landowner must provide a public easement of passage along the shore in return for a building permit, in order to remedy the public's diminished view from the frontal road of

184. Id. at 124.

185. *Keystone Bituminous Coal Ass'n v. DeBenedictis,* 480 U.S. 470 (1987); *First English Evangelical Lutheran Church of Glendale v. County of Los Angeles, California,* 482 U.S. 304 (1987); and *Nollan v. California Coastal Comm'n,* 483 U.S. 825 (1987).

186. *Keystone,* 480 U.S. 470 (1987), involved a state statute requiring that 50 percent of the coal beneath certain structures be left in the ground in order to provide surface support. The Supreme Court found no taking, despite circumstances apparently similar to those in *Pennsylvania Coal.* The Court distinguished that case on two grounds. First, it found significant that *Pennsylvania Coal* had involved a dispute involving private rights (coal companies versus above-ground property owners). Second, the purpose of the act involved in *Keystone* was to protect public safety and to prevent a "nuisance-like" activity—i.e., to prevent damage to public lands and buildings—and not merely to protect the property interests of private parties, as was factually the case in *Pennsylvania Coal.* This second ground supporting the Court's decision solidified the "public safety/nuisance" exception to the notion of regulatory takings—i.e., when the government acts pursuant to its police power to restrict uses of private property that constitute a "nuisance" or "nuisance-like" activity, no taking occurs.

187. *First English,* 482 U.S. 304 (1987). In *First English,* the Supreme Court held that monetary damages are constitutionally required for temporary regulatory takings: "temporary takings which, as here, deny a landowner all use of his property, are not different in kind from permanent takings, for which the Constitution clearly requires compensation." Id. at 318. The appropriate measure of damages is the use value of the land from the time the restriction went into effect until it was rescinded. The Court did not decide whether, as a legitimate exercise of the county's police power to ensure public safety, "the ordinance at issue actually denied appellant all use of its property or whether the county might avoid the conclusion that a compensable taking had occurred by establishing that a denial of all use was insulated as a part of the state's authority to enact safety regulations." Id. at 313 (footnote omitted). On remand, the California Court of Appeals held that the county had acted properly under its police power, that the regulated property retained viable economic use, and consequently that there had been no taking. 210 Cal. App. 3d 1353, 258 Cal. Rptr. 893 (1989).

public beach areas, and the consequent impairment of the public's "psychological" or visual access was a regulatory taking requiring compensation. The Court in *Nollan* added a fourth element to the framework takings analysis set forth in *Penn Central:* to avoid compensation, the government must demonstrate that regulatory exactions (i.e., permit conditions requiring the granting of public rights in private property) "substantially advance" legitimate state interests. The Court was not persuaded by the state's rationale for imposing the permit condition and found that the exaction of a lateral easement across the Nollans' beachfront property bore no relationship to and would not remedy the claimed damage from the Nollans' house construction project to the public's perception that it would gain access to nearby public beach areas.[188]

In July 1992 the Supreme Court decided a challenge brought by a beachfront property owner to a South Carolina statute that prevented him from constructing habitable structures on his property.[189] The statute prohibited construction seaward of a setback line established upon the basis of erosion data collected over a forty-year period. In the majority opinion's view, this setback limit effectively deprived the owner of all economic use of his property. The South Carolina Supreme Court in its decision did not reach the issue of any residual economic value of the property, after imposition of the setback limit, holding that the legislature had properly determined that construction seaward of the setback line would be harmful to public interests, and based upon substantial United States Supreme Court precedents,[190] the court found no taking when the government acts to prevent a "nuisance-like" activity (i.e., the "nuisance exception").

Reversing the South Carolina court's decision, the Supreme Court significantly modified takings analysis by establishing a new *per se* rule that a taking occurs whenever the government by regulation deprives an owner of all economic value of his or her property. The so-called nuisance exception to a takings finding was limited to nuisances that courts have established in the past and that are recognized as nuisances in common law. The authority of the legislature to recognize new and to prohibit or restrict additional "nuisance-like" activities, regardless of the degree of harm to the public, was sharply curtailed. In what has already proved to be a controversial and troubling decision, the conservative majority attacked a long-standing practice of the Court in deferring to the legislative branch in determining matters of public policy (e.g., defining public interests and harms to those interests):

188. *Nollan,* 483 U.S. 825. It should be noted that although the *Nollan* case involved access to the shore, the case was not argued on behalf of the state on public trust principles but on police power theory alone, stressing the state's authority to impose restrictions on development permits in order to protect public interests.

189. *Lucas v. South Carolina Coastal Council,* 112 S.Ct. 2886, 120 L.Ed.2d 798 (1992).

190. These precedents included *Mugler v. Kansas,* 123 U.S. 623 (1887), *Penn Central,* 438 U.S. 104 (1978), and *Keystone,* 480 U.S. 479 (1987).

Any limitation so severe [prohibiting "all economically beneficial use of land"] cannot be newly legislated or decreed (without compensation), but must inhere in the title [to the land] itself, in the restrictions that background principles of the State's law of property and nuisance already place upon land ownership. A law or decree with such an effect must, in other words, do no more that duplicate the result that could have been achieved in the courts—by adjacent landowners (or other uniquely affected persons) under the State's law of private nuisance, or by the State under its complementary power to abate nuisances that affect the public generally, or otherwise.[191]

In effect, judges, rather than legislators, will decide what activities conducted by private property owners may harm public interests. The new, categorical rule that regulation prohibiting a harm recognized by the legislature (the branch of government constitutionally empowered to protect public welfare and interests) requires compensation if economic use of the land is cut off will be administered solely by judges who must discern whether "background principles of nuisance and property" allow government regulation. The majority opinion proposes some guidance for judges exercising this extraordinary power:

The "total taking" inquiry we require today will ordinarily entail . . . analysis of, among other things, the degree of harm to public lands and resources, or adjacent private property, posed by the claimant's proposed activities, [all citations omitted] the social value of the claimant's activities and their suitability to the locality in question, and the relative ease with which the alleged harm can be avoided through measures taken by the claimant and the government (or adjacent private landowners) alike. The fact that a particular use has long been engaged in by similarly situated owners ordinarily imports a lack of any common-law prohibition (though changed circumstances or new knowledge may make what was previously permissible no longer so). So also does the fact that other landowners, similarly situated, are permitted to continue the use denied to the claimant.[192]

The Court's analysis in *Lucas* emphasizes the role of common law nuisance doctrine in deciding whether government regulation of private property restricting all economic uses of the property requires compensation. The majority opinion, however, also invokes "background principles" of property law as justi-

---

191. *Lucas,* 112 S.Ct. 2886, 2900.

Justice Blackmun, dissenting, objected to the Court's "reliance on common-law principles of nuisance" in balancing public and private interests:

In determining what is a nuisance at common law, state courts make exactly the decision that the Court finds so troubling when made by the South Carolina General Assembly today: they determine whether the use is harmful. Common-law public and private nuisance law is simply a determination whether a particular use causes harm. [citations omitted] There is nothing magical in the reasoning of judges long dead. They determined a harm in the same way as state judges and legislatures do today. If judges in the 18th and 19th centuries can distinguish a harm from a benefit, why can not judges in the 20th century, and if judges can, why not legislators? There simply is no reason to believe that new interpretations of the hoary common law nuisance doctrine will be particularly "objective" or "value-free." [citation omitted] Once one abandons the level of generality of *sic utere tuo ut alienum non laedas* [citation omitted], one searches in vain, I think, for anything resembling a principle in the common law of nuisance. *Lucas,* at 2914.

192. *Lucas,* at 2901.

fication for regulation depriving a property owner of all economic use of land. This basis for an exception to the Court's new categorical taking rule can, without question, be supplied by public trust principles. In fact, there can be relatively few more traditional common law property principles than the public trust doctrine. In light of the analysis of this doctrine contained in this study, and considering the potentially significantly limiting effects of the *Lucas* decision upon regulatory takings law generally, the public trust doctrine as a basis upon which to regulate the use of privately held property subject to the doctrine assumes an even greater importance for coastal management. In contrast to the uncertainties inherent in reliance upon nuisance law as a basis for regulation, as Justice Blackmun warns, the public trust doctrine provides rational, coherent principles to guide the management of lands, waters, and resources subject to the doctrine, as argued in this book. Although it is difficult to predict the manner in which a Supreme Court might react to this application of its decision in *Lucas,* nevertheless the Court itself has given coastal managers an additional and substantial reason to ground their management actions in public trust principles.

2. Regulating Private Activities on or Affecting Public Trust Areas

A state's assertion or reassertion of the state's public trust property interests, unlike state regulatory action based solely upon the state's traditional police power, should not be vulnerable to takings challenges. Public trust property rights have been held by the governmental trustee in trust or subject to the trust from the outset and therefore nothing has been taken when the trustee acts pursuant to such rights. Thus, in terms of the framework takings analysis briefly discussed above, a private owner cannot have any reasonable investment-backed expectations with respect to private uses of trust property that are inconsistent with or harmful to the public's trust interests in the property.[193] In *Mono Lake,* the California Supreme Court explicitly addressed this issue and refused to find a taking: "Once again we reject the claim that establishment of the public trust constituted a taking of property for which compensation was required: 'we do not divest anyone of title to property; the consequence of our decision will be only that some landowners . . . hold [title] subject to the public trust' " (citing *City of Berkeley,* 26 Cal.3d 515, 532).[194]

---

193. See discussion of *Penn Central* above.

194. *Mono Lake,* 189 Cal. Rptr. at 359, 658 P.2d at 722. See also *McDonald v. Halvorson,* 92 Or. App. 478, 760 P.2d 263 (1988), in which the court, having established in an earlier case that "dry sands" areas along the Oregon coast are subject to the public's right of recreational use as a matter of law, refused to find a taking in applying that doctrine to private property, finding that the private property owner had failed to show that it had a protected property interest; See also *Boston Waterfront Development Corp. v. Commonwealth,* 378 Mass. 629, 393 N.E.2d 356 (1979).

The California Supreme Court's finding in *Mono Lake* that a state's assertion of its public trust property interest is not vulnerable to a takings challenge finds strong support in a series of cases over almost one hundred years (of which the following are illustrative) that have considered the interaction between the public trust doctrine and takings law. For example, in *Frost v. Railroad Co.*,[195] a littoral landowner's access to navigable waters was blocked by the construction of a bridge across the mouth of an inlet. The Maine court found no taking on the ground that the landowner had no private right of access to these waters but only shared in the public right:

> The sovereign had the absolute control of it and could regulate, enlarge, limit or even destroy it . . . and this without making or providing for any compensation to such individuals as might be inconvenienced or damaged thereby. . . . If, in the evolution of life and commerce, the sovereign comes to believe that the public good will be increased by the creation of some new or additional means of communication and commerce at the expense or even sacrifice of some older one enjoyed merely as a public right, the sovereign can so ordain, even to the detriment of individuals.[196]

In *Xidis v. City of Gulfport*,[197] the Mississippi Supreme Court upheld the right of the trustee to build a small-craft commercial harbor in navigable waters which hindered the access to water of an adjoining riparian owner without paying compensation: "[T]his Court has definitely held that the soil below high tide mark underlying the waters of Mississippi Sound . . . is owned by the State as trustee for its people . . . and that the rights of riparian owners are subject to the prior right of the State to impose an additional public use upon lands without requiring the payment of additional compensation."[198]

Whether the state must compensate a private landowner for limiting or prohibiting uses that are permissible under the public trust doctrine is considered in the Washington case *Orion Corp. v. State*.[199] The Orion Corporation owned

195. 96 Me. 76 (1901).

196. Id. at 85–86.

197. 72 So.2d 153 (1954); see also *Crary v. State Highway Comm'n*, 68 So.2d 368 (Miss. 1953) (riparian's right to plant and gather oysters and construct buildings subordinate to state's public trust interests); *State v. City of Cleveland*, 74 N.E.2d 438 (Ohio 1947) (city may build road blocking riparian's access to navigable waters); *Miramar Co. v. City of Santa Barbara*, 143 P.2d 1 (Cal. 1943) (city may construct breakwater reducing flow of sand to littoral owner's property causing erosion and migration of the mean high-tide line inland); *Colberg, Inc. v. State*, 432 P.2d 2 (Cal. 1967) (state may build bridge blocking plaintiff's access to navigable water and deny more that 80 percent of plaintiff's current business); *in accord, Game and Fresh Water Fish Comm'n v. Lake Islands*, 407 So.2d 189 (Fla. 1982); *Nelson v. De Long*, 7 N.W.2d 342 (Minn. 1942); *State v. Masheter*, 203 N.E.2d 325 (Ohio 1964); *Grupe v. California Coastal Comm'n*, 212 Cal. Rptr. 578 (Cal. App. 1 Dist. 1985); *Whaler's Village Club v. California Coastal Comm'n*, 220 Cal. Rptr. 2 (Cal. App. 2 Dist. 1985); *Wilson v. Commonwealth*, 583 N.E. 894 (Mass. App. Ct. 1992).

198. Id. at 158, quoting *Crary v. State Highway Comm'n*, 68 So.2d 468.

199. 109 Wash.2d 621, 747 P.2d 1062 (1987), *cert. den'd*, 486 U.S. 1022 (1988).

tidelands in Padilla Bay which it planned to dredge and fill to build a residential community. After Orion had purchased the land, the state adopted a series of coastal and tidelands regulations, the effect of which was to limit Orion's use of its property to nonintensive recreation or aquacultural uses.

The Orion Corporation challenged the new coastal and tideland regulations as regulatory takings. The Washington Supreme Court held that title to Washington's shorelines vested in the state when the state entered the Union[200] and that therefore when Orion purchased its tideland property, it did so subject to the public trust. Thus Orion could have no reasonable investment-backed expectations other than those permissible under the public trust doctrine. The court also noted, however, that some of the uses of the land which would be permissible under the public trust doctrine may not be permissible under the challenged tidelands regulations. The court stated that, to the extent the challenged regulations prohibited uses which would violate the public trust, they would be insulated from a takings claim. However, to the extent the regulations prohibited uses which were not impermissible under the public trust, and to the extent those regulations, though enacted for the public health and safety, denied Orion Corporation all economically viable uses of the property, the regulations would constitute a taking.[201]

The Washington court added that if there were a regulatory taking, the state would be given the choice of two remedies: it could either pay compensation for the taking or amend its regulations to permit the Orion Corporation to use the property for such uses as would be allowable under the public trust doctrine.[202] Any takings compensation would be limited to the reduction in value of the land as a result of the new regulations, taking into account that the value of the land was limited from the outset by restrictions inherent in the public trust doctrine.

Thus states may bolster their efforts to protect public trust areas by defining and regulating permissible uses under the public trust doctrine with a high degree of certainty that such regulation will survive a takings challenge. However, if the state trustee has not established reasonably specific uses applicable in public trust areas,[203] and imposes regulations that are more restrictive than are justified by the public trust doctrine in that state, such regulation may be challenged as a regulatory taking.

200. See ch. 1, sec. C for discussion of the equal footing doctrine.

201. 109 Wash.2d at 654, 747 P.2d at 1090.

202. If the state chose to amend its regulations, the Orion Corporation would be entitled to compensation for the temporary regulatory taking in the amount of the leasehold value of the land for the period while the regulations were in effect. Id. at 668, 747 P.2d at 1088.

203. For example, through appropriate legislation; such uses may also be recognized by state courts under existing state public trust doctrine principles.

3. Expanding the Geographic Scope of the Doctrine and Providing Access to Trust Lands

Perhaps the more interesting issue relating to the interaction of the public trust and takings doctrines concerns the risk faced by states that seek to expand the geographical scope of the public trust doctrine or to provide access to public trust areas across private property.

As discussed in ch. 2, sec. A, the geographic extent of the public trust doctrine is traditionally limited to the lands subject to the tides, state coastal waters, and navigable fresh waters. Recent cases, however, suggest that states may extend their public trust authority beyond its traditional boundaries in order to protect public trust areas and resources from injury arising out of activities conducted outside public trust areas. In California, for example, public trust authority extends to control the diversion of water from non-navigable tributaries to Mono Lake, a public trust body of water.[204] As noted above, the plaintiffs successfully argued in *Mono Lake* that under the public trust doctrine the state must consider the effects of diverting water from non-navigable tributaries of Mono Lake upon the lake and its wildlife resources before permitting the diversion of water to southern California.[205] The *Mono Lake* decision is an important precedent for California and other states seeking to extend their authority, under the public trust doctrine, to lands above the traditional boundary in order to provide adequate protection to public trust areas and resources from activities on such lands.

Another approach, endorsed by the New Jersey courts, is to extend the public trust boundary itself to include the public use of privately owned dry sand areas and to permit access across private property to public trust beaches. As stated by the New Jersey Supreme Court:

> The bathers' right in the upland sands is not limited to passage. Reasonable enjoyment of the foreshore and the sea cannot be realized unless some enjoyment of the dry sand area is also allowed. The complete pleasure of swimming must be accompanied by intermittent periods of rest and relaxation beyond the waters' edge. The unavailability of the physical situs for such rest and relaxation would seriously curtail and in many situations eliminate the right to recreational use of the ocean. . . . [W]here use of dry sand is essential or reasonably necessary for enjoyment of the ocean, the [public trust] doctrine warrants the public's use of the upland dry sand area subject to an accommodation of the interests of the owner.[206]

204. *Mono Lake,* 33 Cal.3d 419, 189 Cal. Rptr. 346, 658 P.2d 709.
205. Id.
206. *Matthews v. Bay Head Improvement Ass'n,* 471 A.2d at 365 (N.J. 1984). See also *Lusardi v. Curtis Point Property Owners Ass'n,* 86 N.J. 217, 430 A.2d 881 (1981); *Van Ness v. Borough of Deal,* 78 N.J. 174, 393 A.2d 571 (1978); *Hyland v. Borough of Allenhurst,* 148 N.J. Super. 437, 372 A.2d 1133 (1977); *Borough of Neptune City v. Borough of Avon-by-the-Sea,* 61 N.J. 296, 294 A.2d 47, 51 (1972).

Based upon this rationale, the New Jersey courts have consistently struck down attempts by local communities and private property owners to close off access to their beaches.

In contrast to the New Jersey holdings, the Maine Supreme Judicial Court has recently denied access to beaches subject to public rights under the doctrine across privately held land for purely recreational (as opposed to traditional public trust) purposes. In *Wells Beach*,[207] private landowners challenged Maine's Public Trust and Intertidal Land Act, which provided in part that the public's trust rights included "the right to use intertidal land for recreation" and as a "footway between points along the shore." A narrow majority (4–3) of the Maine Supreme Judicial Court found the act an unconstitutional violation of the takings clause of the United States Constitution, explicitly refusing to expand the common law public trust principles. Thus, at least in Maine, a state statute attempting to enhance public access to public trust resources may not survive a takings challenge.[208]

Before attempting to provide access to public trust areas or to protect public trust resources, however, a state must consider the prospective risk that its action will be challenged as a regulatory taking, whatever the source of its authority. The most recent example of this risk is illustrated by the Supreme Court's decision in *Nollan*, discussed above.[209] *Nollan* was not argued on public trust grounds, although the case arose from the California Coastal Commission's attempt to provide public access across private beachfront property in order to create a passageway linking two public beaches. The United States Supreme Court held that the requirement that the Nollans dedicate a public easement across their property (a strip of land between the mean high-tide line and a seawall) constituted a regulatory taking. In so holding, the Supreme Court emphasized the character of the governmental action as a physical occupation.

A regulation or restriction that results in the public's "physical occupation" or use of land, or any restrictions on the landowner's right to exclude others from the property, is clearly the type of regulation most susceptible to a takings challenge. Traditionally, the right to exclude others from one's property has been deemed a "fundamental stick" in the bundle of property rights.[210]

207. 557 A.2d 168 (1989).

208. The Maine court reaffirmed that private owners of shorefront property held title subject to an easement permitting public use for fishing, fowling, and navigation (whether for recreation or business) and other related uses. 557 A.2d 168 (1989). Consequently, a statute guaranteeing access across private property for such purposes would presumably have been upheld.

209. 483 U.S. 825.

210. A striking example of the effect of this right is found in the Supreme Court's decision in *Loretto v. Teleprompter Manhattan CATV Corp.*, 458 U.S. 419 (1982). In the *Loretto* case, a New York statute required landlords of rental property to permit cable television companies to install cable

In the public trust context, coastal managers face a risk in requiring private owners to grant, without compensation, interests in property (i.e., easements, rights-of-way) for the purpose of providing public access to the beach or shore. The holding in the *Nollan* case is instructive. The Supreme Court found that a land use regulation must "substantially advance legitimate state interests," and that, "unless the permit condition serves the same governmental purpose as the development ban, the building restriction is not a valid regulation of land use but an 'out and out plan of extortion.' "[211] Because the condition imposed by the commission did not contribute toward resolving the problem created or worsened by the proposed development (blockage of visual access), the condition was ruled an unconstitutional taking.

The majority opinion in *Nollan*, however, suggests that the state would be permitted to require access over the Nollans' beachfront property if the state could establish that such a condition would "substantially advance a legitimate state interest" and if the condition itself would mitigate the adverse effect of the Nollans' building project upon such state interest (i.e., by obstructing access to public trust beaches). Thus it is possible to read the *Nollan* opinion as opening the door for public access across beachfront property to public trust beaches if the state can demonstrate that the property owner's project requiring a state permit impairs the state interest in public access to such areas.

The dissenting opinion in *Nollan* supports this interpretation of the Court's ruling and observes that state officials should, without great difficulty, be able to satisfy the higher level judicial scrutiny of such conditions which must now "substantially advance" legitimate public interests. The more direct these relationships, the easier it will be for the state to demonstrate compliance with this new standard. If the state has met this standard, according to the dissenting opinion, a court will then engage in the analysis required by *Penn Central*, examining: the economic impact of the condition or regulation on the property; effects on the property owner's reasonable "investment-backed expectations"; and the character of the governmental action. The *Nollan* dissent suggests that, where a regulation or condition is found to "substantially advance" a legitimate

---

facilities on their property in return for a one dollar payment. Although the physical intrusion was minimal (two small boxes and a cable on the roof), the court found a taking. In doing so, the court adopted a per se rule that "a permanent physical occupation authorized by government is a taking without regard to the public interests it may serve." 458 U.S. at 426. The court went on to state: "In short, when the 'character of the governmental action' is a permanent physical occupation of property, our cases uniformly have found a taking to the extent of the occupation, without regard to whether the action achieves an important public benefit or has only minimal economic impact on the owners." 458 U.S. at 434–35 (quoting *Penn Central*, 438 U.S. at 124).

211. *Nollan*, 483 U.S. at 837 [quoting *J.E.D. Assoc., Inc. v. Atkinson*, 121 N.H. 881, 584, 432 A.2d 12, 14–15 (1981)].

public interest (e.g., access to public trust lands and waters), the "physical oc-cupation" of the property by the public required by the condition will not be fatal, so long as the other elements of the *Penn Central* analysis do not support a takings finding. Thus, if states argue that securing public access to public trust beaches is a legitimate public interest, then conditions which have a direct rela-tionship to the activity or project subject to the permit and which "substantially advance" the legitimate public interest should not be found to be takings even though they may constitute a permanent physical occupation of private property by the public.

The Role of the Public Trust Doctrine in Coastal Management

Twenty years after the passage of the federal Coastal Zone Management Act (CZMA), coastal managers are reassessing their programs to manage coastal areas and resources. A number of factors have contributed to this reassessment, including increasing pressure from coastal development, a growing recognition of the fragility of coastal environments, more accurate estimations of the economic value of coastal areas and resources, advances in the natural sciences providing a much more vivid picture of the impacts of human activity on coastal ecology, serious concern about the deteriorating quality of coastal waters, and a judicial evisceration of the legal tools and management programs to protect these areas and resources during the 1980s and early 1990s.[1] In short, coastal management is changing, and we are entering a period in which more comprehensive coastal management policies and strategies will be necessary to protect the coastal zone and its resources.

The purpose of this part is to examine selectively how coastal states currently use public trust principles in coastal management; to suggest a range of actions that coastal managers should consider to make more effective use of the doctrine; and to analyze in detail how public trust principles affect certain regulatory areas (e.g., access to public trust beaches, permitting private wharves and docks, tidelands oil and gas leasing, etc.).

A. THE USE OF PUBLIC TRUST PRINCIPLES IN COASTAL MANAGEMENT PROGRAMS

This section summarizes our review of coastal state constitutions, statutes, and regulatory programs that explicitly incorporate the public trust doc-

1. The national coastal zone management program was under attack throughout the two Reagan administrations, which attempted to eliminate the program and succeeded in significantly cutting its funding. See Archer and Knecht, 15 Coastal Management 103 (1987). The decisions of an increasingly conservative judiciary denuded local, state, and federal governments of the power to regulate uses of private property that adversely affect public values by requiring the payment of compensation under the takings clause of the Constitution. See e.g., *Lucas*, 112 S.Ct. 2886, 120 L.Ed.2d 798 (1992).

trine, incorporate public trust principles without explicitly incorporating the public trust doctrine, or do not incorporate public trust principles or the public trust doctrine. The purpose of this analysis is to determine whether and how states are using the public trust doctrine and public trust principles to advance their goals of comprehensive management over trust lands and resources.

As discussed in this section, many states use public trust principles and have incorporated them into their coastal management programs. However, this section recommends that states and state agencies can and should do more to incorporate explicitly the public trust doctrine and public trust principles into their constitutions, statutes, coastal zone programs, and general coastal regulations. Incorporation of the public trust doctrine and principles will facilitate comprehensive coastal planning and management and will provide support to state statutes and regulatory programs in the face of potential takings challenges.

The use of public trust principles in coastal management is neither new nor infrequent. With the development of formal coastal management programs under the CZMA, coastal states took on greater responsibility for protecting and managing coastal resources and uses. Some state programs explicitly incorporate and use public trust principles effectively in carrying out this responsibility. Other state programs characterize trust responsibilities in terms that are similar to public trust principles.[2] A few state programs make little use of public trust principles.

Our review of selected state programs suggests that public trust principles are an increasingly common and critical factor in coastal zone management. Several sources of state authority are examined below for expressions of public trust principles either currently or potentially applicable to coastal management: state constitutions, coastal zone management program documents, and coastal management practice in the states.

1. State Constitutions

Several coastal states have constitutional provisions that, although they often do not use the term *public trust*, clearly recognize the responsibilities of the state to manage and preserve its public trust lands and resources. For example, Florida incorporates a traditional public trust approach in its constitution, prohibiting sale or private use of "trust" lands except in the public interest.[3]

2. For example, the Washington State Supreme Court has held that the state Shoreline Management Act is reflective of public trust principles even though the act does not formally acknowledge the public trust doctrine. See *Orion Corp v. State*, 109 Wash.2d 621, 747 P.2d 1062 (1987), *cert. den'd,* 486 U.S. 1022 (1988).

3. Fla. Const. art. 10, sec. 1: "The title to lands under navigable waters, within the boundaries of

Other states have taken a more expansive view of their trust responsibilities, extending it beyond traditional public trust resources. For example, the North Carolina Constitution establishes a policy of conservation and protection of state lands and waters for the benefit of all citizens and recognizes that the state's wetlands, estuaries, and beaches are part of its "common heritage."[4] The Hawaii Constitution recognizes the application of public trust principles in the management of all its natural resources: "For the benefit of present and future generations, the state and its political subdivisions shall conserve and protect Hawaii's natural beauty and all natural resources, including land, water, air, minerals and energy sources, and shall promote the development and utilization of these resources in a manner consistent with their conservation. . . . All public natural resources are held in trust by the state for the benefit of the people."[5] The same article provides that the state accepts an "obligation to protect, control and regulate the use of Hawaii's water resources for the benefit of its people."

In 1979 Virginia adopted a new constitution which states "that it shall be the policy of the Commonwealth . . . to protect its atmosphere, lands and waters from pollution, impairment or destruction, for the benefit, enjoyment and general welfare of the people of the Commonwealth"[6] and that "the natural oyster beds, rocks and shoals in the waters of the Commonwealth shall not be leased, rented or sold but shall be held in trust for the benefit of the people of the Commonwealth."[7]

In 1971 Pennsylvania amended its constitution by adding an "environmental rights" provision: "The people have a right to clean air, pure water, and to the preservation of the natural, scenic, historic and aesthetic values of the environment. Pennsylvania's public natural resources are the common property of all the people, including generations yet to come. As trustee of these resources, the Commonwealth shall conserve and maintain them for the benefit of all the people."[8]

The 1963 Michigan Constitution states: "The conservation and development of the natural resources of the State are hereby declared to be of paramount

---

the state, which have not been alienated, including beaches below mean high water lines, is held by the State, by virtue of its sovereignty, in trust for all of the people. Sale of such lands may be authorized by law, but only when in the public interest. Private use of portions of such land may be authorized by law, but only when not contrary to the public interest."

4. N.C. Const. art. 14, sec. 5. These "common heritage" rights are further refined in the language in N.C.G.S. sec. 146 et seq. For example, N.C.G.S. sec 146–6(f) (1985) states that title to lands raised along the Atlantic Ocean by publicly financed projects vests in the state and will be open to the citizens "consistent with the public trust rights in ocean beaches, which rights are part of the common heritage of the people of this State."

5. Haw. Const. art. 11, sec. 1.

6. Va. Const. art. 11, sec. 1.

7. Va. Const. art. 11, sec. 6.

8. Pa. Const. art. 1, sec. 27.

public concern in the interest of the health, safety and general welfare of the people. The Legislature shall provide for the protection of the air, water and other resources of the State from pollution, impairment and destruction."[9]

California has taken an unusual approach by specifically providing in its constitution for public access to public trust lands:

> No individual, partnership, or corporation, claiming or possessing the frontage or tidal lands of a harbor, bay, inlet, estuary, or other navigable water in this State, shall be permitted to exclude the right of way to such water whenever it is required for any public purpose, nor to destroy or obstruct the free navigation of such water; and the Legislature shall enact such laws as will give the most liberal construction to this provision, so that access to the navigable waters of this State shall be always attainable for the people thereof.[10]

The need for legislation to implement this type of constitutional provision varies from state to state. In Pennsylvania, for example, the courts have ruled that its constitutional amendment described above is "self-executing" and creates citizens' rights and Commonwealth duties.[11] In other words, even though the legislature may pass legislation defining more explicitly the meaning of the constitutional language, no such action is necessary to create citizens' rights to bring individual cases. In Hawaii, citizens' rights are authorized explicitly by the constitution.[12] More commonly, however, the state constitutional recognition of public rights with respect to the shore and other natural resources is not self-executing and must be carried into effect by legislation.[13]

2. Coastal Zone Management Programs and State Law

Although public trust concepts have been incorporated into several state constitutions, as illustrated above, it is more common for states to articulate public trust principles in their coastal zone management legislation and federally approved program documents.[14] Virtually all coastal states have recognized the

---

9. Mich. Const. art. 4, sec. 52.

10. Cal. Const. art. 10, sec. 4.

11. *Pennsylvania Gas and Water Co. v. Kassab*, 14 Pa. Cmwlth. 564, 322 A.2d 775 (1974). *Payne v. Kassab*, 11 Pa. Cmwlth. 14, 312 A.2d 86 (1973), *aff'd*, 468 Pa. 226, 361 A.2d 263 (1976).

12. Haw. Const. art. 11, sec. 9: "Each person has the right to a clean and healthful environment, as defined by law relating to environmental quality, including control of pollution and conservation, protection and enhancement of natural resources. Any person may enforce this right against any other party, public or private, through appropriate legal proceedings, subject to reasonable limitation and regulation as provided by law."

13. E.g., *State v. Cozzens*, 2 R.I. 561 (1850), construing article 1, section 17 of the Rhode Island Constitution; Amend. art. 49 of the constitution of Massachusetts.

14. Ohio, which does not have a federally approved coastal program, nonetheless recognizes the historical uses of navigation, water commerce, and fishing. Further, Ohio law states that Lake Erie and its submerged lands belong "to the state as proprietor in trust for the people of the state, for the public uses to which it may be adapted." Ohio Rev. Code Ann. sec. 123.03.

importance of coastal areas to their economic and environmental well-being. State coastal program documents emphasize the richness and vitality of coastal systems and affirm a role for government in protecting ecologically sensitive coastal and marine areas.

Some coastal state programs clearly incorporate public trust duties and principles, whereas other programs contain public trust-like provisions but do not specifically refer to the doctrine. For example, the North Carolina Coastal Area Management Act states as one of its goals the "[p]rotection of present common-law and statutory public rights in the lands and waters of the coastal area."[15] In contrast, the goals set forth in Washington's Shoreline Management Act of 1971 do not explicitly mention the public trust doctrine, although they have been found to reflect public trust principles.[16]

In Massachusetts, public trust principles are codified in chapter 91 of the Massachusetts General Laws, which provide broad management authority to the Department of Environmental Protection over the conservation and preservation of the public's rights in tidelands. The purpose of the regulations under chapter 91 is to protect the public interest in "tidelands, Great Ponds, and non-tidal rivers and streams in accordance with the public trust doctrine as established by common law and codified in the colonial Ordinances of 1641 and 1647 and subsequent statutes and case law of Massachusetts."[17] In Connecticut, a public trust is declared "in the air, water and other natural resources of the State of Connecticut and . . . each person is entitled to the same protection, preservation and enhancement of the same."[18] Maine's Shoreland Zoning Law[19] states that the state trustee must manage "shoreland areas" and water "to promote public health, safety and the general welfare."

15. N.C.G.S. secs. 113A-102(b)(4)(f) (1983).

16. Wash. Rev. Code sec. 90.58.020 (1971):

> The legislature finds that the shorelines of the state are among the most valuable and fragile of its natural resources and that there is great concern throughout the state relating to their utilization, protection, restoration and preservation. . . . The legislature further finds that much of the shorelines of the state and uplands adjacent thereto are in private ownership; that unrestricted construction on the privately owned or publicly owned shorelines of the state is not in the best public interest; and therefore, coordinated planning is necessary in order to protect the public interest associated with the shorelines of the state. . . .
> It is the policy of the state to provide for the management of the shorelines of the state by planning for and fostering all reasonable and appropriate uses. This policy is designed to ensure the development of these shorelines in a manner which, while allowing for limited reduction of rights of the public in the navigable waters, will promote and enhance the public interest.

The Supreme Court of Washington has held that the Shoreline Management Act is reflective of public trust principles, even though they are not explicitly articulated in the act. See *Orion,* 109 Wash.2d at 640–41, 747 P.2d at 1072–73.

17. 310 C.M.R. sec. 9.01(2)(a); see ch. 5 for a detailed discussion of chapter 91.

18. Conn. Gen. Stat. 22a-15 (1971). Although the Connecticut program goes beyond the traditional public trust doctrine's application to lands beneath navigable and tidal waters, it clearly expresses the principles of the public trust doctrine.

19. Me. Rev. Stat. Ann. tit. 38, sec. 435 (Supp. 1989).

The examples cited above demonstrate that many coastal state programs already incorporate public trust principles either explicitly or implicitly. These principles, however, are not consistently employed by coastal managers, and their acknowledgment in state program documents and statutes does not necessarily lead to effective coastal management based upon the public trust doctrine, in part because of the relative lack of appreciation of the scope of the doctrine. Coastal management practices that have been relatively successful in incorporating public trust principles are discussed below.

3. Management Practices and Public Trust Principles

A number of coastal states have developed several management practices either based upon or reflective of public trust principles, including: designating special areas for public trust management (e.g., areas of particular concern) and areawide management (e.g., the San Francisco Bay Conservation and Development Commission), setting priorities for public trust uses in the coastal zone, and utilizing local government authority. Each practice is considered below.

   a. Special Area Designations and Areawide Management. All state coastal zone management programs identify at least one or more "areas of particular concern" meriting special management, in part because the program development provisions of the CZMA[20] require states to identify such areas before they can receive federal approval and funding for their coastal programs. Coastal managers in some states have tended to undervalue the usefulness of designating "areas of particular concern" and have made only limited use of this coastal management technique, at least as the technique is described in the CZMA. But all coastal programs have engaged in special area management to some degree and some states have effectively used special area designations to protect valuable public trust lands and resources.

Many states have designated special protection zones for areas of particular ecological interest. Typically, no or limited development is allowed in such zones, and management is carried out directly by the state. For example, Florida has designated an "aquatic preserves system," a "state wilderness system," and a system of "state conservation and recreation lands." These are areas owned and managed directly by the state, with the aid of local and regional government representatives. In response to Florida's growth over recent years, the state has further identified "areas of critical state concern" as special management areas.[21] Within these areas the state and local governments have the lead responsibility to

20. 16 U.S.C. secs. 1454, 1455.
21. Three such areas presently exist: Big Cypress Swamp in Collier, Monroe, and Dade counties; the Florida Keys in Monroe County; and the Green Swamp in the west-central part of the state. Florida Coastal Management Program, part 2, sec. 2(c)(3).

develop regulations that comply with state policies guiding and limiting development in these areas. Under the Florida program no more than 5 percent of the area of the state may be identified as critical areas.

In certain states, however, the strategy has been to classify large sections of public trust lands and waters within different use categories, subject to different regulatory controls. For example, North Carolina has designated large areas of its public trust lands and resources as "areas of environmental concern" (AECs) within which coastal permits are required for development activities.[22] North Carolina grouped all its coastal areas into broad management categories.[23] According to state law, these areas constitute "an interrelated group of Areas of Environmental Concern, so as to safeguard and perpetuate their biological, social, economic, and aesthetic values and to ensure that development occurring within these [areas] is compatible with natural characteristics."[24] The AECs are interdependent and require management as a unit embracing "all characteristics, process, and features of the whole system and not characteriz[ing] individually any one component of an estuary."

Rhode Island, whose program includes public trust-like principles but makes little effective use of the doctrine in actual coastal management, divides all state waters into six management categories designed to help "maintain a high quality of coastal environment for future generations of Rhode Islanders." These waters are categorized according to their location and present use. The state has also developed a set of use criteria designed to protect and/or enhance current uses within each type of coastal environment.[25] Rhode Island has also identified

22. North Carolina's AECs include:

All waters of the Atlantic Ocean and the lands thereunder from the mean high water mark to the seaward limit of state jurisdiction; all natural bodies of water subject to measurable lunar tides and lands thereunder to the mean high water mark; all navigable natural bodies of water and lands thereunder to the mean high water level or mean water level as the case may be, except privately owned lakes to which the public has no right of access; all water in artificially created bodies of water containing significant public fishing resources or other public resources which are accessible to the public by navigation from bodies of water in which the public has rights of navigation; and all waters in artificially created bodies of water in which the public has acquired rights by prescription, custom, usage, dedication, or any other means. N.C. Admin. Code tit. 15r. sec. 7H.0203 (Amend. 1988).

23. For example, within the estuarine system, the program identifies four categories of environmental concern: estuarine waters, coastal wetlands, public trust areas, and estuarine shorelines. Id.

24. Id.

25. Rhode Island Coastal Resources Management Program, sec. 200.1:

Type 1 waters abut shorelines in a natural and undisturbed condition, where alterations, including the construction of docks and any dredging, are considered . . . unsuitable.

Type 2 waters are adjacent to predominantly residential areas, where docks are acceptable, but more intense forms of development, including marinas and new dredging projects (but not maintenance dredging), would change the area's character and alter the established balance among uses. Alterations such as these would bring more intensive uses and are therefore prohibited in Type 2 waters.

Type 3 waters are dominated by commercial facilities that support recreational boating. Here,

certain areas that require more detailed attention, for which they have developed a series of "special area management plans."[26] In practice the Rhode Island approach has meant: the assigning of special management and review authority to a subcomittee of the state's coastal management agency (the Coastal Resources Management Council); the development of a cooperative review process for all involved state agencies; and the creation of an "action committee" for coordinating further planning, education programs, and other nonregulatory initiatives.[27] One of the reasons articulated by Rhode Island for developing a special area management strategy is that: "[a]gencies of state and local government . . . engaged in the review process grant permits in a sequential, usually isolated manner. This [procedure] reduces the integration of the diverse concerns of individual agencies. While the decisions reached in this manner may be legally valid, they forego the opportunity to increase their effectiveness."[28]

The several state programs utilizing special area designations vary considerably in their use of public trust principles. North Carolina, for example, bases its designation approach directly on its public trust authority. In contrast, Rhode Island does not incorporate the doctrine directly into its special area management plans, although its classifications describe public trust-like principles in

---

marinas, boatyards, and associated businesses take priority over other uses, and dredging and shoreline alterations are to be expected.

Type 4 waters include the open waters of the Bay and the Sounds, where a balance must be maintained among fishing, recreational boating, and commercial traffic. Here high water quality and a healthy ecosystem are primary concerns.

The last two water use categories [type 5 and type 6] are assigned to areas adjacent to ports and industrial waterfronts. In these waters, maintenance of adequate water depths is essential, high water quality is seldom achievable, and some filling may be desirable. Within type 5 ports, a mix of commercial and recreational activities must coexist, whereas in type 6 waters water-dependent industrial and commercial activities take precedence over all other activities.

26. Illustrative of this approach is Rhode Island's Narrow River Special Area Management Plan. The strategy behind the plan "involves recognition of all components involved in a specific ecosystem and the complexity of interactions which have evolved among these components. Subsequently, the disturbance or alteration of just one component of the system can have far-reaching effects, often unexpected and occasionally irreversible." The Narrow River Special Area Management Plan, sec. 110.A (1987).

Goals for the plan include:

1. To provide for a balance of compatible uses, consistent with [the state's] responsibility for preserving, protecting, and restoring coastal resources; specifically, to guide the actions of private citizens, municipalities and state agencies in the restoration and maintenance of environmental quality in the Narrow River;

2. To provide a regional plan for the Narrow River that recognizes that the watershed functions as an ecosystem . . . ; and

3. To create a decision-making process appropriate to the management of the watershed as an ecosystem, specifically insuring consideration of long-term cumulative impacts. Id. at sec. 140.A 1–3.

27. Id. at sec. 220.1.

28. Id. at sec. 210.3.A.

some detail. Explicitly stating public trust principles as the bases or reasons for coastal management legislation, coastal agency regulations, and coastal agency adjudicatory decisions (including permitting) is one of the most effective means of invigorating and legitimizing the public trust doctrine.[29]

Another coastal management technique based in part upon public trust principles has been the creation of management authorities for large, well-delineated bodies of water or public trust regions. Areawide management has a number of advantages. First, it permits states to identify and define areas based upon their actual geographic features and natural resource values, rather than on artificial municipal or county borders, and to develop comprehensive management plans targeted to address the concerns specific to that region. Second, where a large body of water or public trust region spans more than one state, areawide management may provide a structure for states to work together to develop comprehensive planning and management strategies for their common resources.[30] Third, areawide management programs may be structured, using the public trust doctrine and the state's police powers in tandem, to encourage comprehensive management over the array of lands and resources within the area, thus avoiding the limitations inherent in ad hoc permitting decisions.[31]

An interesting example of an areawide management plan is the San Francisco Bay Conservation and Development Commission (BCDC). Although BCDC has been in existence since 1965 and predates the CZMA, BCDC is also part of the California Coastal Management Program. The act that created the commission required the preparation of "a comprehensive and enforceable plan for the conservation of the water of San Francisco Bay and the development of its shoreline."[32] In general, BCDC's statutory authority coincides with and is reflective of the state's authority as trustee of the public trust lands and resources of San Francisco Bay.[33]

29. See ch. 2, secs. D and G.

30. See, as an example, the Chesapeake Bay Commission of Maryland and Virginia. Md. Code Ann. sec. 8-302; Va. Code Ann. secs. 62.1-69–62.1-69.20. See also ch. 2, sec. G.

31. A similar technique has achieved some prominence recently through the development of comprehensive estuary management plans by states participating in the U.S. Environmental Protection Agency's National Estuary Program under the Clean Water Act.

32. McAteer-Petris Act, Cal. Gov. Code 66600 (West 1983).

33. Specifically, under its statutory mandate, BCDC:

Regulates all filling and dredging in San Francisco Bay (which includes San Pablo and Suisun Bays, sloughs and certain creeks and tributaries that are part of the Bay system, salt ponds, and certain other areas that have been diked-off from the Bay);
Preserves the Suisun Marsh, the largest remaining wetland in California, by administering the Suisun Marsh Preservation Act in cooperation with local governments;
Regulates new development within the first 100 feet inland from the Bay to ensure that maximum feasible public access to the Bay is provided;
Minimizes pressures to fill the Bay by ensuring that the limited amount of shoreline area

In carrying out these tasks, BCDC has instituted several useful strategies. For example, the commission has implemented an aggressive preapplication assistance program to aid developers in coordinating and acquiring the necessary permits and meeting the provisions of the bay plan. In this preapplication review, the commission is assisted by three volunteer professional boards that review the design, engineering, and scientific/technical aspects of projects.[34] Any dredging, filling, or any substantial change in bay or shoreline uses (including repairs or improvements to existing structures) requires a commission permit. One of the fundamental issues addressed by the BCDC in its permitting process is the question of filling by developers of the shallow bay waters. Under BCDC direction, new coastal development has contributed a total of 760 acres of new water surface to the bay since 1970 (374 acres have been filled while 1,134 acres of new bay water have been opened).[35]

b. Prioritizing Public Trust Uses. A persistent and serious problem confronting coastal managers has been to regulate competing uses of public trust lands and resources. Although allowing certain uses and prohibiting other uses within an area may be done under the police power of a state or local government, such regulation may be vulnerable to takings claims under the Constitution.[36] The public trust doctrine, however, permits coastal states to prioritize uses of public trust lands and resources by restricting or prohibiting uses that are detrimental to the trust and only permitting consistent uses, and by creating a ranking among permissible uses, with less risk of claims of unconstitutional takings.[37]

Coastal state programs have developed use priorities for public trust lands and resources, in part because the federal CZMA required them to do so in developing their programs.[38] For example, Washington's coastal zone program

---

suitable for high priority water oriented uses is reserved for these purposes. These priority use areas include: ports, water-related industries, water-oriented recreation areas, airports and wild-life areas;

Pursues an active planning program to study Bay issues so that commission plans and policies are based upon the best available current information; and

Administers the federal Coastal Zone Management Act with the San Francisco Bay segment of the California coastal zone to ensure that federal activities reflect the Commission's policies. 1988 Annual Report, San Francisco BCDC, p. 1.

34. The Design Review Board is comprised of prominent design and engineering experts who advise on the appearance, design, and public access of proposed projects. The Engineering Criteria Review Board uses science and engineering specialists to review projects that involve fill or which pose problems relating to the safety of structures on filled lands. The Scientific and Technical Advisory Committee advises the commission on the accuracy and completeness of scientific information used to evaluate proposed bay development. Id. at 2.

35. Id. at 5.

36. See ch. 2, sec. H for discussion of the takings doctrine.

37. See ch. 2, sec. B for discussion of appropriate and protected public trust uses and the difficult issue of choosing among competing uses. See also ch. 2, secs. D and G for discussion of the breadth of coastal management agency discretion to make such choices.

38. 16 U.S.C. sec. 1454.

requires its Department of Ecology, in adopting guidelines for "shorelines of state-wide significance," to give preference to uses that: "(1) recognize and protect state-wide interests over local interests; (2) preserve the natural character of the shoreline; (3) result in long-term over short-term benefit; (4) protect the resources and ecology of the shoreline; (5) increase public access to publicly owned areas of the shorelines; and (6) increase recreational opportunities for the public in the shoreline."[39] Such a list may not offer detailed guidance to coastal managers, but it illustrates a general effort to evaluate the appropriateness of competing uses in a public trust area and provides a statutory basis for disapproving certain uses while allowing others.

Connecticut has addressed the question of use regulation within the context of an overall resource management/impact zoning approach used by both state and municipal agencies. Each applicant for a development permit must undertake a multistep evaluation to determine the degree to which a proposal is consistent with the coastal policies of the state. First, the applicant must identify the coastal resources on or near the proposed activity that may be affected, which will, in turn, determine which of the state's coastal and water resource policies are applicable to the project. Second, the applicant must review the coastal use policies to determine their applicability. Finally, before seeking a permit from municipal zoning and state regulatory programs, the applicant must assess the potential adverse impacts of the project on the affected coastal resources.[40]

Connecticut has designated twelve separate categories of coastal land and water resources and developed discrete use guidelines for all activities carried out within these designated areas.[41] Within each area category, the state has also created policies for prioritizing uses, as well as descriptions of "activities and uses generally consistent" with applicable coastal policies.[42] An applicant must

39. Wash. Rev. Code sec. 90.58.020 (1971).

40. Such adverse impacts include:

Degrading water quality through the significant introduction into either coastal waters or groundwater supplies of suspended solids, nutrients, toxics, heavy metals or pathogens, or through the significant alteration of temperature, pH, dissolved oxygen or salinity;

Degrading or destroying essential wildlife, finfish or shellfish habitat through significant alteration of the composition, migration patterns, distribution, breeding or other population characteristics or the natural species or significant alterations of the natural components of the habitat; and

Degrading tidal wetlands, beaches and dunes, rocky shorefronts and bluffs and escarpments through significant alteration of their natural characteristics or function. The burden of demonstrating consistency with state coastal policies lies with the applicant. Conn. Gen. Stat. sec. 22a-106 and sec. 22a-93(15).

41. These twelve categories are bluffs and escarpments, rocky shorefronts, beaches and dunes, intertidal flats, freshwater wetlands and watercourses, coastal hazard areas, developed shorefront, islands, shorelands, shellfish concentration areas, and coastal waters and estuarine embayments Conn. Gen. Stat. sec. 22a-93(7).

42. Id. For example, within the defined "beaches and dunes" category, the state has identified several use policies to preserve natural beach systems:

A. To preserve the dynamic form and integrity of natural beach systems in order to provide

not only demonstrate that its proposed project is consistent with the state's coastal policies and will not lead to adverse environmental impacts, but also that the proposal is consistent with the use guidelines for activities at the specific category.

Another example of priority categories has been developed in Rhode Island. Although Rhode Island does not specifically rely upon the public trust doctrine in regulating activities and projects in its coastal zone, the state nonetheless employs a use classification system that reflects public trust principles. For each of its coastal water classification categories (described in the previous section), Rhode Island has devised a matrix setting out the level and intensity of review for uses within each category.

North Carolina has one of the more highly developed and detailed criteria for evaluating uses of public trust lands and resources. It establishes both general and specific use standards for public trust areas. For example, within "estuarine waters," the state has asserted that the: "[h]ighest priority of use shall be allocated to the conservation of estuarine waters and its vital components. Second priority of estuarine waters use shall be given to those types of development activities that require water access and use which cannot function elsewhere such as simple access channels; structures to prevent erosion; navigation channels; boat docks, marinas, piers, wharfs, and mooring pilings."[43] Within public trust lands: "[i]n the absence of overriding public benefit, any use which significantly interferes with the public right of navigation or other public trust rights which the public may be found to have in these areas shall not be allowed."[44] These general evaluative criteria are bolstered by lengthy sets of specific standards to regulate activities within public trust areas.[45]

c. Utilizing Local Government Authority. Some coastal states have delegated to local governments the responsibility to make use decisions affecting public trust lands and resources, typically on the theory that local agencies have a greater familiarity with local conditions. For example, in Washington, local master programs are a primary mechanism for coastal zone management decision making. Local authorities develop these programs under guidelines set forth in the state's Shoreline Management Act. Upon approval by the Washington De-

---

critical wildlife habitats, a reservoir for sand supply, a buffer for coastal flooding and erosion, and valuable recreational opportunities;

   B. To insure that coastal uses are compatible with the capabilities of the system and do not unreasonably interfere with natural processes of erosion and sedimentation; and

   C. To encourage the restoration and enhancement of disturbed or modified beach systems. Conn. Gen. Stat. sec. 22a-90 et seq.

43. N.C. Admin. Code tit. 15r. sec. 7H.0206(d) (Amend. 1988).
44. N.C. Admin. Code tit. 15r. sec. 7H.0207(d) (1977).
45. Id.

partment of Ecology, local master programs regulate uses of the various shore-lines of the state.[46] To date, 217 local master programs have been approved.

As an example, the Skagit County Shoreline Management Program sets out guidelines for permitted uses as follows: "permitted uses in the shorelines of the state shall be designed and conducted in a manner to minimize, insofar as practical, any resultant damage to the ecology and environment of shoreline areas and any interference with the public's use of the water."[47] The Skagit County program creates six shoreline area designations intended to "provide a systematic, rational, and equitable basis upon which to guide and regulate development within shoreline areas."[48] Within each of these area designations, the master program defines a discrete set of management objectives and policies designed to protect certain resources and uses and to limit or restrict specific kinds of activities.[49] The management policies in such areas provide for compatible, appropriate uses that do not conflict with natural and cultural processes and features of the water body and its associated wetlands. Such uses are required to be shoreline and water dependent. Further, "[p]riority should be given to those activities which create the least environmental impact to this shoreline area."[50] Skagit County, like Rhode Island, has developed an "allowable use matrix" to guide planners in regulating uses within all areas governed by the master program.

Massachusetts, Alaska, and California are additional examples of a move toward local authority. Massachusetts has recently implemented a program to aid in the development of local harbor plans. The commonwealth has prepared a set of guidelines to be used by local authorities and provided funds and technical and legal assistance to direct the preparation of such plans.[51]

46. Wash. Rev. Code Ann. sec. 90.58.100 (1971).

47. Skagit County Shoreline Management Master Program, sec. 103 (1983) (cited hereafter as Skagit County Program).

48. The six designations under the Skagit County program are: urban shoreline area, rural residential shoreline area, rural shoreline area, conservancy shoreline area, natural shoreline area, and aquatic shoreline area. Skagit County Program, sec. 6.1.

49. For example, an "aquatic shoreline area" is defined to include "all state water bodies." This area designation includes state marine waters, lakes, and rivers together with their underlying lands and water column, including bays, straits, harbor areas, waterways, coves, estuaries, lakes, stream-ways, tidelands, bedland, and shorelands. Skagit County Program, secs. 6–12, 13. The objective of such a designation is "intended to encourage and protect appropriate multiple uses of the water or, in some cases, single purpose, dominant uses in limited areas; to manage and protect the limited water surfaces and foreshores from inappropriate activities or encroachment."

50. Skagit County Program secs. 6–13.

51. These guidelines articulate numerous reasons for the preparation of local harbor plans, including the finding that "[m]any communities face deteriorating environmental quality as they attempt to accommodate the demands of competing harbor uses. Through harbor planning, communities will be able to analyze cumulative environmental impacts and develop Proposals to limit

One of the goals of the Alaska Coastal Management Program (ACMP) has been to strengthen local planning in order to "furnish coastal area citizens with improved opportunities to constructively influence the land and water management decisions which affect their lives."[52] The ACMP establishes "coastal resource districts" as its basic local governmental units. To date, thirty-three local coastal district programs have been developed and approved by the Alaska Coastal Policy Council.

In California, local communities are mandated by the state to develop "local coastal plans," which serve as a primary focus for coastal management and planning. As in other jurisdictions, the state does not transfer complete authority to local government. Rather, the California Coastal Commission retains original permit jurisdiction over development on "tidelands, submerged land, [and] on public trust lands, whether filled or unfilled, lying within the coastal zone."[53] The state also retains appellate jurisdiction over certain locally issued coastal permits, including those involving development in the immediate vicinity of the ocean shoreline, beaches, estuaries, wetland, streams, and "any sensitive coastal resource area."[54]

Other states tend to limit the authority of local jurisdictions to regulate public trust areas. For example, in Louisiana, local governments may develop local coastal management programs, but both state and local coastal managers are restricted in their authority to design comprehensive management plans under the state program. Activities relating to "the location, drilling, exploration, and production of oil, gas, sulfur or other minerals" do not require coastal use permits at either the state or local level.[55] Additionally, several other classes of activities are exempt from the coastal use permitting process.[56] Such exemptions from the permitting process obviously limit the ability of state and local authorities to manage the impacts of development activities on coastal resources coherently.

---

those effects. Local boards and agencies can better utilize state and local environmental standards by tailoring by-laws or new programs to meet local needs identified in the planning process." Mass. Office of Coastal Zone Management, Harbor Planning Guidelines, p. 2 (1987).

52. Alaska Coastal Management Program, ch. 3, sec. a.

53. Cal. Pub. Res. Code sec. 30519(b) (West 1982).

54. Cal. Pub. Res. Code sec. 30603 (West 1982).

55. La. Rev. Stat. Ann. sec. 49:213.12 (West 1978).

56. These include: "(1) Agriculture, forestry, and agriculture activities on lands consistently used in the past for such activities; (2) Normal maintenance or repair of existing structures, including emergency repairs of damage caused by accident, fire, or the elements; and (3) Construction of a residence or camp." La. Rev. Stat. Ann. sec. 49:213.15 (West Amend. 1983).

B. USING THE PUBLIC TRUST DOCTRINE MORE EFFECTIVELY

This section focuses on strategies for states and coastal managers to make more effective use of the public trust doctrine and public trust principles in their constitutions, statutes, coastal zone programs, and general coastal regulations: incorporating public trust principles into comprehensive areawide management; establishing priorities among competing public trust uses and specifying prohibited uses of public trust lands and resources; and addressing the serious problem of cumulative effects of unrelated coastal development projects. This section also recommends that legislators and coastal managers should invoke the public trust doctrine and public trust principles in their existing coastal decision-making processes in order to strengthen their legal authority and to insulate their regulatory actions from constitutional takings challenges.

This section considers a number of strategies that coastal managers may employ to bring public trust principles directly into coastal management decision making. This discussion is framed by two major cautionary observations. First, although there is a common core content to the public trust doctrine, the scope of the doctrine varies from state to state. This variation requires that readers first carefully review the status of the doctrine in their state in order to evaluate the viability of any given action to be taken under public trust principles. Second, state coastal programs also vary greatly in their treatment of the doctrine. Some state programs make effective use of public trust principles in coastal management. Not surprisingly, these programs are found in states where the doctrine is relatively robust. Some state programs make little use of the doctrine in support of coastal management, either because the doctrine's uses are not understood to extend beyond the traditional uses of navigation, fishing, and commerce or because coastal managers are unfamiliar with public trust principles.

A middle group of state programs could, with relative ease, make highly effective use of the doctrine by invoking and applying its principles in coastal management planning and decision making. As demonstrated in the earlier discussion of coastal programs, public trust criteria may be found in state program documents and policies, even where there is little or no acknowledgment of the public trust doctrine. Through fairly routine implementation of existing coastal programs, by revising current program regulations, or in some cases, by amending state programs, public trust principles can be introduced into coastal management planning and decision making in a conscious, deliberate manner.

Our consideration of strategies to bring public trust principles more effectively into coastal management is directed primarily at the second and third categories of state programs described above. Our concern is not so much with specific actions to be taken by coastal managers in certain states (although such actions would spring from these strategies), but with a range of general strategies. These proposals seek to maximize certain features of the doctrine (such as the scope of authority given to state trustees over trust property to prioritize trust uses) that can be employed effectively in conjunction with police power authority (such as land use planning and regulation) to enhance a coastal agency's ability to manage coastal resources and uses.

This section examines first the advantages of comprehensive areawide management combining public trust principles and police power. This strategy would encompass most if not all of the goals of coastal zone management, including preserving coastal features such as beaches, dunes, estuaries, and wetlands, ensuring public access to trust lands and resources, and prohibiting or restricting nontrust uses of coastal resources. Second, we consider the use of public trust principles in addressing the difficult problem of managing the cumulative effects of a multitude of uses in the coastal zone. Finally, we discuss the obvious but nonetheless substantial advantages accruing to coastal managers by the relatively simple act of working public trust principles into coastal management practice.

1. Comprehensive Areawide Management

As stated above, there are compelling environmental reasons to manage related resources within a specific area comprehensively. Briefly, uses of certain resources within the area have inevitable effects upon other resources and uses. Governing such uses requires that they be evaluated comparatively from both ecological and management perspectives, in order to balance use and protection of resources and resolve conflicts among competing uses. Coastal management is itself areawide management of the coastal zone and its resources.

The overlap between the coastal zone and the "zone" of public trust resources and interests, however, allows states to ground their management programs in public trust principles as a source of legal authority in addition to the state's police power. There are two major advantages to this strategy of combining public trust principles with traditional police power: first, obtaining legal justification deriving from public trust property principles for both prioritizing and prohibiting uses within the public trust/coastal management area; and second, providing a defense of management decisions from regulatory takings claims, in addition to the protection afforded such decisions under the state's police power.[57]

a. Prioritizing and Prohibiting Uses. The previous section cited examples of

57. See discussion of takings law, ch. 2, sec. H.

the practice of state coastal programs in prioritizing uses of public trust and coastal lands and resources. Such rank orders of uses may be justified as an exercise of inherent state police power, but are subject to regulatory taking claims that may limit the extent to which coastal states may employ a priority system for coastal resources and uses.[58] The public trust doctrine, however, has permitted different priorities among traditional uses at different times. Historically, the doctrine recognized a priority for navigation over other uses of public trust lands.[59] But the doctrine has accommodated changing priorities over time, and navigation no longer necessarily enjoys its former premier status.

It has been argued that public trust theory is deficient because the doctrine does not itself impose a hierarchy of uses of public trust lands and resources. A more useful view is that the doctrine's flexibility permits states to strike a different weighing of social values and a different balance of uses at different times. This flexibility and expansiveness can be seen in the growing number of uses that courts (and coastal managers) have found to be subject to the public trust doctrine.[60] For example, access to public trust beaches has become a public trust use of high priority in some states (e.g., access is protected in California by the state constitution;[61] New Jersey's courts have elevated beach access to priority status.)[62]

The public trust doctrine justifies distinguishing between permissible trust uses and impermissible nontrust uses. Here the doctrine's effect is clearest—impermissible uses can be prohibited by the trustee with little risk of successful taking claims.[63] Further, the state may establish an order among permissible public trust uses recognized by state law. As demonstrated above, states such as North Carolina and Connecticut have ranked types of public trust lands and waters and identified the types of uses and activities that may occur in each category. Use priorities are more likely to be upheld in the face of claims of a regulatory taking if the priority is established as part of a comprehensive plan of management adopted by the state identifying specific important public interests and explicitly grounding choices in public trust principles.

58. Id.

59. See Note, "The Public Trust in Tidal Areas: A Sometimes Submerged Traditional Doctrine," 79 Yale L. J. 762, 777–87 (1969–70). An ordering of priorities is necessary between the different rights as well as vis-à-vis other claims. The different claimed rights have different social weights. Thus the right to unencumbered navigation will usually assume a higher value than the right to collect seaweed on the shore. The weights can be debated, and they will change as social conditions change; but they are indisputably not all equal.

60. See ch. 2, sec. B.

61. Cal. Const. art. 20, sec. 4.

62. See ch. 2, sec. B and ch. 3, sec. C, subsection 1.

63. See discussion of *Orion* decision in ch. 2, sec. H.

b. The Public Trust and Cumulative Effects. Addressing the problems caused by the cumulative effects of numerous projects has been extremely difficult for coastal managers. We have suggested earlier that public trust principles, based on the trustee's property interest in trust lands and resources, in conjunction with traditional state police power, provide a strong and legally supportable ground for areawide management and prioritization of coastal uses. This management framework also allows state trustees to set limits upon the cumulative effects of large numbers of individual projects affecting trust lands, waters, and uses. For example, limits may be imposed on new trust uses on the theory that such uses damage the public interest in other uses of trust lands and resources. Recognizing the cumulative effects problem and acting on its obligation to preserve trust property, a state may decide that an increased rate of development threatens this public interest and must be limited to prevent such harm.[64] Further, the state trustee may justify a limit upon individual uses or projects because of their cumulative effects on trust lands and resources, on the ground that no trust use may be permitted without adequate consideration of its effect on other trust uses.[65] Although disappointed applicants may challenge such limits as regulatory takings, a properly applied and articulated public trust doctrine should provide an effective shield against such claims.[66]

2. Identifying Public Trust Authority in Coastal Programs

Coastal managers could make more effective use of public trust principles if such principles were explicitly contained in coastal programs. Examples of coastal programs that have in fact incorporated elements of the doctrine have been provided. In several of these programs, public trust lands and waters have been identified and allowable uses specified. States that have not categorized public trust lands and resources or specified and prioritized uses within such categories based upon public trust principles should take steps to do so. Even programs that have already identified coastal (and public trust) areas as subject to special management within the coastal zone should consider whether an improved and detailed justification for special areawide management could be established within the program. The state's purpose should be to clarify the relationship between

64. See discussion of *Illinois Central* and conveyancing in ch. 2, sec. F, discussing conveyancing standard that conveyance not cause harm to remaining trust land; also *see* discussions of trustee's fiduciary duties in ch. 3, sec. C (duty to preserve trust property) and of the judicial deference typically shown toward coastal agency decisions which have been conscientiously and carefully made, in ch. 2, secs. D and G.

65. This approach draws support from the *Mono Lake* case, in which a California court held that the state's public trust authority extended to restrict diversions of upstream waters in order to prevent negative impacts on downstream public trust resources. See ch. 2, sec. A.

66. See ch. 2, sec. H for discussion of takings.

actions taken to manage such lands and resources and public trust principles, in order to ground coastal use permitting decisions securely within the public trust doctrine.

Coastal states may consider a range of actions to incorporate public trust principles into their coastal programs, including routine program changes, rule making under existing law to establish land and water use categories and priorities, or new legislation to implement the common law doctrine. For example, where coastal programs already incorporate public trust principles in program authorities, managers may issue regulatory guidelines for planning and permitting actions that fully describe the relevant public trust authority. Rule making may be required for programs that incorporate basic public trust authority but lack any comprehensive trust land and water use categories to guide planning and permitting decisions. Where state programs do not incorporate such basic public trust authority, state agencies may initiate new legislation.[67]

3. Invoking Public Trust Principles

Effective use of public trust principles requires that coastal managers systematically invoke such principles in their planning and permitting decisions. *Post hoc* rationalizations of *ad hoc* permitting actions claiming public trust authority are often ineffective in defeating legal challenges. Although this recommendation may seem obvious, coastal managers do not uniformly make use of the public trust doctrine to justify permitting decisions affecting trust lands and resources. If the doctrine is not used as support for a state agency decision (either in permitting or regulation), reviewing courts may not consider the state's public trust authority in reviewing the decision, increasing the likelihood that the court will fail to understand and recognize the public interests protected by the doctrine, and find that the state has acted improperly or has taken private property without just compensation. As discussed in chapter 2, sections D, G, and H, in addressing regulatory actions involving trust lands and resources, the public trust doctrine may be used effectively to establish the public interests and purposes served by regulatory conditions and restrictions, thereby insulating such conditions from claims of a regulatory taking.

C. APPLICATION OF THE PUBLIC TRUST DOCTRINE TO SELECTED ISSUES IN COASTAL MANAGEMENT

Having reviewed general public trust principles and the role that they may play in coastal management, this section examines in more detail seven areas

---

67. See generally ch. 2, sec. D.

of special concern to coastal management: (1) beach access and beach fees; (2) erosion control; (3) private docks and wharves; (4) oil and gas development in tidelands; (5) aquaculture; (6) environmental protection and estuaries; and (7) harbor development and maintenance.

1. Beach Access and Beach Fees

The United States is rapidly approaching a crisis in the availability of beaches that are open to the public. Traditionally, states and citizens have taken one of three approaches to provide access to public trust lands to the public. First, if a path to the beach has been used by the members of the public over a sufficient period of time, it may be considered a public easement under the related common law doctrines of custom, implied dedication, or prescription.[68] There are, however, evidentiary problems with the theory of easements, involving proof of public use as well as acquiescence by private landowners.

Second, the state may take an easement from the property owner by eminent domain. Such takings require compensation to the property owner under the Fifth and Fourteenth Amendments to the United States Constitution (as well as equivalent state constitutional provisions), and could be prohibitively expensive as a means of acquiring public access to beaches.

Third, as a valid exercise of its police power, the state may regulate the use of privately owned lands adjacent to public trust beach areas and require that the owners of such lands grant easements of access to the public when such owners seek permission to use their property in some fashion (e.g., construction or reconstruction projects requiring local and state permits). Such regulatory exactions, however, may be vulnerable to takings claims unless the state can demonstrate an appropriate relationship or nexus between the access easement requirement and the public interest which the easement serves.[69]

---

68. At common law, these were distinct doctrines with different elements. For example, implied dedication required donative intent on the part of the property owner as well as acceptance by the donee, while prescription was limited to cases of adverse interest—the user gained the right of use against the interest and without the permission of the owner. In modern cases, however, several courts have broadly interpreted the doctrines, finding sufficient evidence of donative intent in an owner's failure to object to use [*Seaway Co. v. Attorney General*, 375 S.W.2d 923 (Tx. 1964)], and prescription where public use has existed for five years or more [*Gion v. City of Santa Cruz*, 2 Cal.3d 29, 84 Cal. Rptr. 162, 465 P.2d 50 (1962)]. But see *Bell v. Town of Wells*, 557 A.2d 168, 179 (Me. 1989) ("very few American states recognize the English doctrine of public easements by local custom"). Connecticut rejected this doctrine in *Graham v. Walker*, 78 Conn. 130, 133–34, 61A. 98, 99 (1905).

69. The required nexus between the easement exacted from the owner would probably be provided, or at least substantially strengthened, by reliance upon the public trust doctrine. See ch. 2, sec. H for discussion of regulatory takings and the public trust doctrine and a discussion of *Nollan v. California Coastal Comm'n*, 483 U.S. 825 (1987). According to the majority opinion in *Nollan*, "land use regulation does not effect a taking if it 'substantially advance[s] legitimate state interests' and does not 'den[y]' an owner economically viable use of his land.' " Id. at 834 [quoting *Agins v. Tiburon*, 447

An alternative to these three traditional approaches, depending upon the law of the specific state, would be to use the public trust doctrine and constitutional statements of public trust principles to open access to public trust lands. If the public trust doctrine is to have substance, the general public must have reasonable access to trust lands. If the recreational demands and aesthetic needs of modern society are to be met, it is appropriate for courts, legislatures, and state government agencies to recognize this corollary to the doctrine and move affirmatively to preserve or create such the public right of meaningful access.

One of the critical current issues under the public trust doctrine is the extent to which the public trust includes rights in the public to gain access to public trust lands by passing over privately held lands. States have taken very different approaches to this question. For example, California and New Jersey have assimilated the right of access into the public trust doctrine itself, on the theory that if the trust is to have practical meaning, the public must possess such a right. The New Jersey courts have consistently held that the public must be given access to the dry sand areas as reasonably necessary under the public trust doctrine.[70] California's leading beach access cases were decided before the United States Supreme Court's 1987 decision in *Nollan v. California Coastal Comm'n* and involved challenges to state regulations conditioning issuance of development permits on grants of public easements across private beach property.[71] As in the *Nollan* case, the California Court of Appeals consistently upheld such exactions.[72]

By contrast, the Supreme Judicial Court of Maine recently denied public access across private land to beaches subject to the public trust.[73] Owners of beachfront property sued the town and the state Bureau of Public Lands, among

---

U.S. 255, 260 (1980)]. Legitimate uses of the police power will be upheld even if they are detrimental to the private property owner. Id. at 836–37; *Allied-General Nuclear Services v. United States,* 839 F.2d 1572, 1577 (1988).

70. See ch. 2, sec. A, subsection 2.

71. See ch. 2, sec. H for discussion of *Nollan.*

72. In *Grupe v. California Coastal Comm'n,* 166 Cal. App. 3d 148, 212 Cal. Rptr. 578 (1985), for example, the California Court of Appeals found that such a condition did not violate substantive due process because the condition was reasonably related to a legitimate governmental purpose, i.e., protecting the public's access to tidelands. The court also held that the condition did not affect a taking because the property owner was not deprived of "all" reasonable use and economic value of his property. In *Whaler's Village Club v. California Coastal Comm'n,* 173 Cal. App. 3d 240, 220 Cal. Rptr. 2 (1985), *cert. den'd,* 476 U.S. 1111 (1986), beach homeowners challenged a condition that they dedicate public easements across their properties in order to receive permits to construct a revetment to prevent beach erosion. Again, the court of appeals reversed a trial court decision and found that the revetment met the applicable definition of new development. In so ruling the court noted in dicta that "what respondent disregards is that homeowners in *Whaler's Village* live adjacent to a constitutionally and statutorily designated area owned by the State in trust for all of the people of the State." *Nollan* has been taken by some to raise serious doubts about the validity of these two California rulings, although *Nollan* did not involve or discuss the public trust doctrine.

73. *Wells Beach,* 557 A.2d 168 (1989).

others, in a quiet title action, challenging Maine's Public Trust and Intertidal Land Act,[74] which declared that the public's trust rights included "the right to use intertidal land for recreation" and as a "footway between points along the shore."[75] A bare majority of the Maine Supreme Judicial Court (all of whom reportedly owned beachfront property), over a vigorous and historically correct dissent, narrowly construed public trust rights as limited to the traditional ones of "fowling, fishing and navigation"; refused to extend such rights to include recreational activities; and accordingly found the act an unconstitutional violation of the takings clauses of the Maine and United States constitutions.

The Massachusetts Supreme Judicial Court, because of a somewhat similarly restricted view of the scope of public trust rights beyond those originally recognized, has also found that even a limited recreational easement confined to a path amounted to a permanent occupation requiring compensation under the takings clause.[76] Thus in Massachusetts and Maine there currently exists no access easement, even for walking, above the low watermark and no lateral access to coastal public trust lands, at least for nontraditional public trust activities.

The capacity of public trust principles to justify access to trust lands will be determined by state law. What direction state courts follow will depend on four factors. First, state courts will be guided by their views as to the appropriate role of the judiciary in effecting societal change.[77] Second, the nature and intensity of society's need for recreation will influence the degree to which courts link the public trust doctrine with a right to access public trust lands. In this regard, state courts may be more likely than federal courts to appreciate local opportunities for recreation, budget constraints, and the ramifications of insufficient access to recreation. Third, the doctrine itself will likely evolve with the needs of society, as it has in the past. As noted in the dissent in *Wells Beach*, the doctrine has always been rooted in the public's need for sustenance, by fowling and fishing, and society still needs sustenance, albeit in new and different forms.[78] Finally, courts will be influenced by the degree to which state legislatures pass legislation or constitutional amendments are enacted, declaring that the coast is held in trust

74. 12 M.R.S.A. secs. 571–73 (Supp. 1987–88). Intertidal land is defined in Maine as "land . . . located by the tides between the mean high watermark and either 100 rods seaward from the high watermark or the mean low watermark, whichever is closer to the mean high watermark." *Wells Beach*, 557 A.2d at 169 n. 3.

75. 12 M.R.S.A. secs. 571–73, 573(1)(B) (Supp. 1987–88); § 571.

76. *Opinion of the Justices*, 365 Mass. 681, 313 N.E.2d 561 (1974).

77. Contrast, for example, the *Van Ness* court's admonition that "the situation will not be helped by restrained judicial pronouncements" (78 N.J. at 180, 393 A.2d at 574) with the *Wells Beach* court, "although contemporary public needs for recreation are clearly much broader, the courts and the legislature cannot simply alter . . . long established property rights to accommodate new recreational needs" (557 A.2d at 169).

78. *Wells Beach*, 557 A.2d at 189 (Wathen, J. dissenting).

for the public and that such a trust includes a right of access to the shore. In spite of rulings like *Wells Beach,* such authoritative pronouncements will shift the burden of proof from the state to the private landowners by arguably creating a presumption of public access. Even in states that have constitutional language implying a right of access to public trust lands,[79] the state legislatures should enact implementing legislation to make such access a reality and to provide for its fair and effective implementation.

The courts in at least two states, New Jersey and Florida, have considered the issue of access fees for beach use in light of the public trust doctrine. Courts in both states have held that user fees are permissible so long as they are "reasonable,"[80] the revenue generated from fees is expended for the protection and welfare of the beach and its users,[81] and the fees do not discriminate between residents and nonresidents (at least those who are state citizens).[82]

For example, New Jersey courts have struck down attempts by municipalities to charge higher fees to out-of-town users, stating that "the public trust doctrine dictates that the beach and the ocean water must be open to all on equal terms and without preference and that any contrary state or municipal action is impermissible."[83] New Jersey courts have also consistently struck down indirect attempts to limit access, for example by restricting access to public toilets.[84] The borough of Belmar, New Jersey, and other municipalities have tried to limit access to adjacent beaches by more subtle means. Belmar, for example, charged three dollars for daily weekdays, six dollars for daily weekends, but only forty dollars for a full season. Not surprisingly, Belmar found itself in court, the sub-

79. See e.g., Cal. Const. art. 10, sec. 4. See also, Ak. Const. art. 8, sec. 14; Mich. Const. art. 4, sec. 52; Pa. Const. art. 1, sec. 27; Va. Const. art. 11, secs. 1–3; and Wash. Const. art. 15.

80. See, *City of Dayton Beach Shores v. State,* 483 So.2d 405 (Fla. 1985), and *Borough of Neptune City v. Borough of Avon-by-the-Sea,* 61 N.J. 296, 294 A.2d 47 (1972).

81. In *City of New Smyrna v. Board of Trustees of Internal Improvement Trust Fund,* 543 So.2d 824 (Fla. 1989), the District Court of Appeals of Florida for the Fifth District ruled that beach toll revenue had to be used to keep the beach in a state of repair but did not believe that expenses for beach maintenance must be limited to acts done directly on the sandy beach itself.

82. But see *Broekl v. Chicago Park District,* 131 Ill.2d 79, 544 N.E.2d 792 (1989), in which the court held that where there was clear legislative authority, municipality could charge higher fees to nonresidents for municipal moorings.

83. *Neptune City,* 61 N.J. at 308–9, 294 A.2d at 54. But note that the holding is limited to municipally owned beaches. In *Hyland v. Borough of Allenhurst,* 78 N.J. 190, 393 A.2d 579 (1978), the New Jersey Supreme Court affirmed an appellate division holding that "the general public is entitled to access to both the public trust lands along the Allenhurst shoreline and to all portions of the dedicated beach area in that municipality for a fee no greater than that charged residents for similar use." Id. at 193–94, 393 A.2d at 581 (quoting *Hyland v. Borough of Allenhurst,* 148 N.J. Super. 437, 441, 372 A.2d 1133, 1135 (1977)).

84. *Hyland,* 78 N.J. 190, 393 A.2d 579. Note that although the court struck down the toilet restrictions, it upheld the municipality's right to charge higher fees to nonresidents for use of the beach club facilities.

ject of a suit by the public advocate challenging the beach fee structure.[85] The court held that Belmar is a trustee over its beach area and that it breached its duties and obligations as a trustee by increasing beach fees, rather than real estate taxes, to raise the borough's general revenues.[86] The court also found that Belmar's price structure for beach fees discriminated against nonresidents by imposing a disproportionately high fee on daily and weekend beach badge purchasers.[87]

Parking restrictions could also serve as an effective deterrent to beach use. It remains to be seen, however, whether parking restrictions that are clearly intended to restrict beach access will withstand challenge.[88]

In 1988 a study prepared on behalf of the New Jersey public advocate confirmed what the National Park Service had recognized in its fee structure for a long time: recreational participation is closely associated with affluence. What is surprising is not that courts in New Jersey and Florida have restricted the use of beach fees as a tool to limit beach access, but that there has been so little beach fee litigation elsewhere. As the demand for recreational beaches and appreciation of the public trust doctrine increase, this trend is likely to change.

2. Erosion Control

The point where land and sea meet has always been a dynamic place; therefore, it is not surprising that erosion affects 90 percent of the nation's coasts. The Atlantic Coast erodes an average of two to three feet a year and the Gulf Coast, an average of four to five feet a year.[89] In its natural state, the shore maintains a remarkable equilibrium, and erosion poses little long-term threat to public trust interests such as navigation and fishing. As the shoreline moves landward, so do navigable waters, as well as wetland, dune, and beach areas. Human development, however, often blocks the landward evolution of an eroding shoreline and speeds up the erosion process. This appears true even of most structures built to prevent erosion. Wetlands and beaches disappear, reducing the public's access to and ability to use public trust lands. At the same time, erosion threatens the integrity of shore area structures, inducing the structure's owners to erect shore

---

85. *Slocum v. Belmar,* 569 A.2d 312 (N.J. Super. Ct. 1989).

86. Id. at 317.

87. Id. But see *Broeckl v. Chicago Park District,* 131 Ill.2d 79, 544 N.E.2d 792, upholding higher mooring fees for nonresidents over objections raised under the public trust doctrine.

88. General residential parking restrictions have been upheld by the United States Supreme Court, so long as they "rationally promote" the municipality's objectives. *County Bd. of Virginia v. Richards,* 435 U.S. 5, 7 (1977).

89. N.Y. Times, Aug. 1, 1989.

protection devices which, in turn, typically exacerbate erosion along the neigh-
boring shoreline.[90]

Nonetheless, both states (as the owners of eroding public trust land) and
riparians (as the owners of disappearing upland) typically perceive that they have
an interest in controlling the speed, path, and effects of shoreline erosion. The
battle for erosion control can take three forms: reclamation of lands submerged
due to erosion; preserving the status quo through placement of shore protection
structures and beach nourishment; or retreat by prohibiting shoreline development. To be effective, all of these approaches require cooperation between the
state and riparian owners and implementation on a shorewide rather than a
piecemeal basis. Furthermore, to avoid endless litigation, the relative ownership
rights of the state and riparians as the shoreline moves must be clearly delineated.

Strategies to reclaim submerged lands or prohibit shoreline development
must reflect the realities of a moving shoreline affecting the relative ownership
interests of the state and riparian owners. In most states, accreted and relicted
lands becomes the private property of the riparian owner. Similarly, land submerged by erosion becomes new public trust land. When shoreline changes
result from human actions such as filling or dredging, however, the line of
ownership depends on who caused the change and the purpose behind it. In
general, accreted land resulting from a state project based on public trust purposes becomes public trust land. Similarly, when a riparian's land is eroded at an
increased rate due to a state public trust project, the state owes no compensation
to the riparian for his lost land.

The advantage to the state of adding to its public trust land through reclamation projects may be overshadowed by the costs of such projects, which have led
in some states to statutes that authorize reclamation of submerged lands by
private riparians.[91] For example, Louisiana allows riparians to reclaim lands
submerged due to erosion dating back until 1921.[92] This reclamation must occur
pursuant to a permit issued by the Department of Natural Resources and vests

90. Absent such devices, by the time waves reach the shore they have minimal energy to carry
sand from the shores in their backwash. In contrast, waves hit a sea wall very sharply and powerfully,
building the energy necessary to carry sand away. *Id.* at C12, col. 1. Groins and jetties designed to
keep sand from silting up channels and to trap the sand carried along the shore by currents may
preserve the particular beach, but only at the expense of a beach downstream. Id. at col. 3.

91. "Whereas, in many of the bays, harbors and inlets on the sea coast of this State, the sea is
annually encroaching upon the land, washing away the shores and shoaling such [waters]; and
whereas such encroachments can be prevented only at great expense . . . it is desirable that facilities
and encouragement should be offered to the owners of the soil abutting upon the coast in such
[waters] to make improvements and expenditures that will stay such encroachments." 1872 Or. Laws
p. 129, cited in *Shively v. Bowlby,* 152 U.S. 1, 3 at note 1 (1894).

92. La. Rev. Stat. Ann. sec. 41:1702 (West Supp. 1987).

title in the reclaiming riparian.[93] The public's rights are protected by a prohibition against reclamation that would obstruct navigability or "impose undue or unreasonable restraints on state rights which have vested since the land became submerged."[94] This type of statute can advance the public's interest in lands reclaimed by riparians without burdensome costs to the public. A state could also reserve a public easement for public trust purposes in the reclaimed land, as a condition of each reclamation permit. Such a statutory scheme would benefit public and private interests alike.

Although keeping the shoreline in place is always costly and often technically improvident, it is nonetheless often necessary (even if not legally obligatory) as the only way to protect the investments of countless riparian owners in shore developments as well as to preserve traditional public beach areas. California makes the best of these pressures by limiting shore protection projects to those that serve public interests in the shore. Any construction that alters natural shoreline processes must either serve a coastal-dependent use, protect existing structures, or protect a public beach and be designed to mitigate adverse impacts on the remaining shoreline.[95] Florida also provides an example of how equitably to apportion the ownership interests of a state and riparians when the shoreline changes due to a state shore protection project. Prior to undertaking a shore protection project, the state fixes an erosion control line which represents the mean high-water line of the area to be protected.[96] After a public notice and hearing process, all lands landward of the erosion control line vest in the riparian upland owners.[97] Once this erosion control line is established, the common law rules applicable to changes in the shoreline no longer apply.[98] The riparian and state's ownership interests remain static. This system creates certainty and an incentive for the state and riparians to cooperate on shore protection projects.

As the number of undeveloped shore areas dwindle, the remaining few take on added importance for public trust uses. It is easier to control erosion by preventing disruptive development rather than by mitigating development's adverse erosion impacts. These two factors make construction setbacks in undeveloped shore areas an excellent method of protecting the public's interest in shore areas. Florida's coastal construction setback scheme applies to undevel-

93. Id.
94. Id.
95. Cal. Pub. Res. Code sec. 30235 (West 1989); *see* also Fla. Sta. Ann. sec. 161.061 (West Supp. 1989) (State can require removal of any coastal construction on public trust lands, regardless of date of construction or state authorization, which serves no public purpose in the state's opinion.) Fla. Stat. Ann. sec. 161 (West Supp. 1989).
96. Fla. Stat. Ann. at sec. 161.141.
97. Id. at sec. 161.181.
98. Id.

oped and developed shore areas alike. Counties and municipalities may establish coastal construction zoning and building codes to protect beaches and barrier dunes from erosion by restricting development seaward of a certain point.[99] The state must approve both the substance of these local programs and determine that the locality has adequate funds and personnel to administer the programs.[100] If either of these elements is lacking or a locality otherwise fails to establish such a program, the state may establish a coastal construction control line.[101] Development seaward of the state or local setback line must comply with very restrictive standards.[102] In addition, only single-family dwellings may be erected on land the state determines will be below the high waterline due to erosion within thirty years of the construction permit application date.[103] The state may not take future shore protection, beach restoration, or beach nourishment projects into account in determining this thirty-year erosion level.[104]

Because of each state's proprietary interest in the coastal areas, there is little doubt that it and its duly delegated coastal agencies have the power to undertake the erosion control activities discussed above under its public trust as well as its police power authority. Whether, in the absence of a statute, the state has a duty under the public trust doctrine to reclaim or protect coastal areas altered or threatened by natural forces; and if there is such a duty, whether it is enforceable by private citizens, are questions that do not yet appear to have been addressed, let alone definitively decided, by the courts. Since substantial numbers of beach restoration or nourishment projects have been primarily for recreational enhancement purposes, no such duty would exist with respect to such projects in jurisdictions that had not included recreational activities within the public trust doctrine. Similarly, shore protection projects intended, as so many are, primarily to benefit private residential property owners would not seem compelled by a state's public trust obligation. All coastal erosion control projects must, however,

99. Fla. Stat. Ann. sec. 161.053 (West Supp. 1989) (such programs must adequately "preserve and protect the beaches and coastal barrier dunes adjacent to such beaches . . . from imprudent construction that will jeopardize the stability of the beach-dune system, accelerate erosion, provide inadequate protection to structures, endanger adjacent properties, or interfere with public beach access.") Id. Presumably the local authority should use criteria similar to those used by the state to arrive at a coastal construction line.

100. Id.

101. The state establishes the line based on ground elevations in relation to historical storm and hurricane tides, predicted maximum wave uprush, beach and offshore ground contours, the vegetation line, erosion trends, the dune or bluff line and existing upland development. Id.

102. Seaward development will be allowed only if the impacts of the proposed development on the beach dune system "clearly justify" the development or if the immediately adjacent area is continuously developed seaward of the line and no undue erosion effects have occurred. Id.

103. Id.

104. Id.

be consistent with public trust principles to the extent they are incorporated into the state's coastal zone management plan,[105] as well as satisfy the requirements of federal and state environmental policy acts with respect to assessing and mitigating project impacts upon the environment.

3. Private Docks and Wharves

Courts, legislatures, and managers repeatedly face the same issues with regard to private docks and wharves: the rights of landowners to build private docks, piers, wharves, and piles in the submerged land under the tidewaters; the limitations the state and its designees may place on these rights; and the authority of the state to convey the tidal areas for the erection of water-related structures while subjecting them to scrutiny under the public trust doctrine.

Riparian and littoral owners expect that their property's location on water entitles them to special rights in that water. Common and statutory law have long recognized a riparian's right to build wharves and docks on abutting public trust land in order to gain access to navigable water.[106] Such rights seem to contradict the public trust doctrine and have often been challenged on this ground.[107] In general, courts have tended to uphold such rights to the extent that they can be reasonably construed to promote public purposes (such as enhancing commerce) without substantially impairing the public's interest in or use of trust lands.[108]

It is clear, however, that public trust rights take precedence over conflicting riparian rights.[109] In addition, the federal government's paramount power to regulate navigable waters is superior to both riparian and public trust rights.[110]

---

105. See ch. 4.

106. *Mutual Chem. Co. of America v. Mayor & City Council of Baltimore,* 33 F. Supp. 881, 883 (D.Md. 1940), *aff'd in part and rev'd in part sub nom., Mayor & City Council of Baltimore v. Crown Cork & Seal Co.,* 122 F.2d 385 (4th Cir. 1941); *Sterling v. Jackson,* 69 Mich. 488 (1888); *State v. Shannon,* 36 Ohio St. 423 (1881); *Simpson v. Moorhead,* 56 A. 887 (N.J. 1904).

107. See, e.g., *Caminiti v. Boyle,* 107 Wash.2d 662, 732 P.2d 989 (1987) (plaintiff argued that members of the public "have recreational interests that are affected by their ability to acquire access to and use public aquatic lands and waters. . . . These interests are impacted to some extent by the presence, location, and private use of private recreational docks on these public aquatic lands and waterways.")

108. Id. Cf. *Illinois Cent. R.R. v. Illinois,* 146 U.S. 387 (1892).

109. E.g., *Weeks v. North Carolina Dep't of Natural Resources and Community Development,* 388 S.E.2d 228 (1990). See generally, Cheung, "Dockominiums: An Expansion of Riparian Rights That Violates The Public Trust Doctrine," 16 B.C. Env'l Aff. L. Rev. 821, 841–44 (1989).

110. *Scranton v. Wheeler,* 179 U.S. 141 (1900), 43 U.S.C. sec. 1314(a) (1982) (the "United States retains all its navigable servitude and rights in and powers of regulation and control of said lands and navigable waters for the constitutional purposes of commerce.") The federal government may control navigable waterways for the benefit of the public without compensating adversely affected riparians. *Zabel v. Tabb,* 430 F.2d 199 (5th Cir. 1970), *cert. den'd,* 401 U.S. 910 (1971).

This hierarchy of priorities can create difficult tensions. May a state prohibit the riparian from wharfing out if it is the riparian's only means of access to the water? May a state block a riparian's access by leasing the public trust land lying between the upland and navigable water or by conveying such land to a third party?

These questions have been answered differently by the individual states,[111] because each state determines the nature and extent of riparian rights and the scope of its public trust doctrine.[112] Traditionally, all states recognize a core of riparian rights: the right of contact with and access to the water; the right to use the shore immediately adjacent to the land belonging to the public; the right to have water flow to the owner's land without any artificial obstruction; and the right to erect wharves ("wharf out") to reach the navigable portion of the stream.[113]

The traditional common law right to wharf out does not, in the absence of a statute to the contrary, require prior state approval.[114] This right includes building wharves and piers for commercial use[115] as well as structures serving the personal access needs of the riparian. The right "to wharf out," however, is merely an implied license from the state which exists only as long as the state permits.[116] Landowners who have constructed docks, piers, and other structures may continue to use these facilities unless and until the state determines that their continued existence is inconsistent with the reasonable needs of the public in such land and with public trust interests, particularly with respect to unobstructed navigation and fishing.[117]

When such improvements threaten public rights, a state can either order their removal and reenter the public trust land or restrict the riparian's use of the

111. *Barney v. City of Keokuk*, 94 U.S. 324, 338 (1876).

112. *Shively v. Bowlby*, 152 U.S. 1 (1894).

113. L. H. Farnham, *The Law of Waters & Water Rights*, 62 at 279–80 (1904); *B & O R.R. Co. v. Chase*, 43 Md. 23, 35 A.2d (1875). But see *Weeks v. North Carolina Dep't of Natural Resources and Community Development*, 388 S.E.2d 228 (1990), upholding denial of permit to construct a 900-foot pier to deep water, based in part on the public trust doctrine.

114. *Illinois Cent. R.R. v. Illinois*, 146 U.S. 387 (1892); *Shively v. Bowlby*, 152 U.S. 1 (1894); *Sullivan v. Moreno*, 19 Fla. 200 (1882); *Chicago v. Van Ingen*, 152 Ill. 624, 38 N.E. 894 (Ill. 1894); *Hanford v. St. Paul & D.R. Co.*, 43 Minn. 110, 42 N.W. 596 (1889); *Madison v. Mayers*, 97 Wis. 399, 73 N.W. 43 (1897); and *Brusco Towboat Co. v. State of Oregon*, 30 Or. App. 509 (1977) [citing Farnham, *The Law of Waters & Water Rights*, 278 (1904)].

115. See *Grinels v. Daniel*, 110 Va. 874, 67 S.E. 534, 535 (1910).

116. *Brusco Towboat Co. v. State of Oregon*, 30 Or. App. 509 (1977); *River Development Corp. v. Liberty Corp.*, 51 N.J. Super. 447, 144 A.2d 180 (1958), *aff'd*, 29 N.J. 239, 148 A.2d 721 (1959); *Coquille Mill & Merchantile Co. v. Johnson*, 52 Or. 547, 551, 98 P. 132 (1908).

117. *Coquille*, 52 Or. at 551, 98 P. 132; *Montgomery v. Shaver*, 40 Or. 244, 248, 66 P. 923 (1902); *Bowlby v. Shively*, 22 Or. 410, 420, 30 P. 154, 158 (1892) ("revocable at the pleasure of the legislature until acted upon"), *aff'd*, *Shively v. Bowlby*, 152 U.S. 1, 14 S.Ct. 548, 38 L.Ed. 331 (1894); *State v. Superior Court of Placer County*, 172 Cal. Rptr. 713, 29 Cal.3d 240, 625 P.2d 256 (1981).

improvements. But a state does not have carte blanche under the first option. Although it can eliminate a riparian owner's unexercised right of access based on public trust principles without compensating the riparian,[118] the state may be required to pay just compensation to the riparian evicted from improved public trust land.[119] A state may, however, limit this compensation to the improvements' value as used for the particular purpose authorized by the express or implied authority to construct the improvements.[120] A state may further limit this compensation to the value of the riparian's unrecouped investment in the improvements, rather than the improvements' market value.[121]

Regulating uses avoids the economic limitations associated with a state's re-entry onto improved public trust land. The courts in at least two states have held that a riparian who lawfully wharves out under a legislative grant premised on facilitating shipping cannot later use the public trust land for non-shipping-related purposes. The improvements must be used for those purposes for which the right to wharf out was granted.[122] If, for example, the riparian improved public trust land in the 1800s under a legislative grant in order to provide railroad services to the waterfront, the riparian cannot later use the improved lands for a residential development.[123] This principle of public trust law applies even though a century after the grant is made the original use is no longer feasible.[124] Some state courts have held that the use of the improvements is limited to the specific use for which they were built,[125] whereas other courts would allow any public trust use of the improvements.[126]

Although a state may choose to allow courts to apply common law principles to decide issues of riparians' use of improvements, it may more effectively resolve differences between public trust and riparian rights by affirmatively regulating riparian improvements.[127] By statute, a state may require an express grant or

118. Farnham, *The Law of Waters & Water Rights,* 113b (1904); but see *Game and Fresh Water Fish Comm'n v. Lake Islands,* 407 So.2d 189 (Fla. 1982) (regulation prohibiting use of motorboats on lake during duck hunting season, which effectively denied riparians' access to their property, deprived riparians of their property requiring compensation).

119. Id.; *City of Oakland v. Oakland Water-Front Co.,* 118 Cal. 160, 50 P. 277 (Cal. 1897).

120. Cf. *Orion Corp. v. State of Washington,* 103 Wash.2d 441, 693 P.2d 1369 (Wash. 1987), *cert. den'd,* 108 S. Ct. 1996 (1988) (no reasonable investment-backed expectations in improved public trust lands other than those permissible under public trust doctrine).

121. See e.g., *City of Oakland v. Oakland Water-Front Co.,* 118 Cal. 160, 50 P. 277 (Cal. 1897).

122. *State of Vermont v. Central Vermont Railway,* Docket No. 87-607 (Vt. Sup. Ct. Dec. 22, 1989); *Boston Waterfront Development Corp. v. Commonwealth,* 378 Mass. 629, 649, 393 N.E.2d 356 (1979).

123. *Central Vermont Railway,* Docket No. 87-607 (Vt. Sup. Ct. Dec. 22, 1989).

124. *Id.*

125. Id.

126. *Smith v. New York,* 545 N.Y.S.2d 203 (1989).

127. See, e.g., *Weeks v. North Carolina Dep't of Natural Resources and Community Development,* 388 S.E.2d 228 (1990), upholding denial of a riparian owner's application for a permit to construct a 900-foot pier to reach to deep water, based in part on public trust principles.

other state authorization before riparians may erect new structures or substantially improve existing structures on public trust lands.[128] The riparian has no right to compensation for improvements erected without explicit state approval once the state begins to regulate trust land improvements.[129] Furthermore, the state approval, whether a lease, license, or grant,[130] may be limited to a specific term with neither a right of renewal nor compensation for the improvements once the term expires.[131]

At the same time a state may accord special consideration to riparians under the state's public trust land licensing or leasing program. A riparian who erected or maintained structures on abutting trust lands prior to the effective date of a leasing statute could be granted automatic leases under the statute.[132] Leases for certain small structures, such as private recreational piers accessory to a single-family dwelling, could be granted without compensation to riparians.[133] In New Jersey, riparians have a statutory right of first refusal on all third-party applications for a license to improve public trust land.[134] If the riparian decides not to exercise this right, the grantee or licensee still must compensate the riparian for injury to the riparian's right of access to and use of the water.[135]

Every statutory or regulatory program governing private use of public trust lands should consider including provisions explicitly forbidding state authorization where the proposed wharf or other proposed structure or use would unduly obstruct navigation or interfere with other public trust rights.[136] Dockominiums are one such use that may frequently oust or preempt other public trust activities. A dockominium is a private dock or slip space that an individual owns rather than leases from the state. This private ownership of public trust waters can expunge all public rights in these waters if granted without proper safeguards or conditions. While a state may permanently alienate public trust land in certain situations,[137] the private and potentially exclusive nature of dockominiums argues for particularly close scrutiny before the state consents to any such con-

---

128. A Wisconsin statute regulating wharves, piers, and swimming rafts reflects such legislation: "The wharf or pier is constructed to allow the free movement of water underneath and in a manner which will not cause the formation of land upon the bed of the waterway." Wis. Stat. Ann. sec. 30.13(1) (e) (West 1987); see also, Mass. Gen. L. ch. 91, sec. 23; N.J. Stat. Ann. sec. 12:3–19; Wash. Rev. Code Ann. sec. 79.90.105.

129. Id.

130. See ch. 2, sec. G for discussion of licenses and leases.

131. See, e.g., Mass. Regs. Code tit. 310, sec. 9.00 (Public Hearing Draft, 1989).

132. See, e.g., Ohio Rev. Code Ann. sec. 123.031(D).

133. See, e.g., Cal. Pub. Res. Code sec. 6503.5 (West 1977).

134. N.J. Stat. Ann. sec. 12:3–7 (1979).

135. N.J. Stat. Ann. sec. 12:3–9 (1979).

136. *Keane v. Stetson,* 22 Mass. 492 (1827).

137. See ch. 2, sec. F for discussion of such conveyances.

veyances. Ultimately, no private use of public trust land, including those by a riparian owner, should be unconditionally irrevocable.[138]

4. Oil and Gas Development in State Waters

As the need for domestic sources of energy grows, the coastal states may more frequently consider the public trust doctrine as a source of authority to regulate oil and gas exploration and production within state waters (seaward to the three-nautical-mile limit). This section examines the role of the doctrine in regulating oil and gas development in state waters in Florida, Louisiana, and California.

a. Florida. In an early Florida case,[139] the Supreme Court of Florida was asked to decide who was the owner of minerals lying beneath the beds of navigable waterways below the high watermark. Both the state and the riparian owners asserted exclusive rights to these minerals. After reviewing the history of the public trust doctrine, the court concluded, on the basis of the equal footing doctrine, that the state owned these resources[140] and that these lands and resources were to be available for the use and enjoyment of the people of the state.[141] Moreover, the state, acting as a fiduciary for the public, was required to manage the land and authorize only those uses that would preserve the rights available to public under the doctrine.[142]

b. Louisiana. The right of the state to lease the beds of navigable streams for oil exploration has long been recognized in Louisiana. In *Smith v. Dixie Oil Co.,*[143] the Supreme Court of Louisiana held that the beds of all streams that were navigable at the time the state was admitted to the union were the property of the state below the low-tide line.[144] Therefore, the state may lease the land for exploration and development.[145] However, the state's ownership and the interests of the lessee were subject to the right of the public to navigate the waters above the streambed and to use the stream banks.[146] Therefore, Louisiana, like Florida, may lease public trust land for oil exploration, but the lessee may not interfere with the public's use of the land for public trust purposes.

Louisiana currently has statutory authority for leasing land subject to the

---

138. For a more complete discussion on dockominiums, see Cheung, "Dockominiums: An Expansion of Riparian Rights that Violates the Public Trust Doctrine," 16 B. C. Env'l Aff. L. Rev. 821 (1989). See generally ch. 2, sec. G regarding appropriate conditions on such private use.

139. *State v. Black River Phosphate Co.,* 32 Fla. 82, 13 So. 640 (1893).

140. Id. at 106, 13 So. at 648. See also *State ex rel. Peruvian Phosphate Co. v. Board of Phosphates Commissioners,* 31 Fla. 558, 12 So. 913 (1893); *J. P. Furlong Enterprises, Inc. v. Sun Exploration and Prod. Co.,* 423 N.W.2d 130 (N.D. 1988); *State v. Black Bros.,* 116 Tex. 615, 297 S.W.2d 213 (1927).

141. *Black River,* 32 Fla. at 108–9, 13 So. at 648–49.

142. Id.

143. 156 La. 691, 101 So. 24 (1924).

144. See also *Temple v. Eastham,* 150 La. 247, 90 So. 637 (1922).

145. *Dixie Oil,* 156 La. at 702, 101 So. at 28.

146. Id.

public trust. "[T]he State Mineral Board has authority to lease for the development and production of minerals, oil, and gas, any lands belonging to the state, or the title to which is in the public, including . . . water bottoms."[147] Although there is no provision in the statute stipulating that the lease must be determined to be in the public interest, this omission may merely indicate an assumption by the legislature that oil and gas drilling is necessarily in the public interest.[148] Under section 30:172, the lessee is authorized to construct breakwaters, platforms, islands, and fills upon the shores, banks, or water bottoms. Although such construction may interfere with traditional public trust uses, the potential interference may be mitigated under section 30:172, requiring a lessee to obtain a permit before filling or creating an island in navigable waters.

Despite the lack of an express provision directing that leases be subject to the continued use of the area for public trust purposes, Louisiana law does provide that some of the benefits derived from oil and gas leases further public trust interests. Specifically, section 30:311 creates the Coastal Environment Protection Trust Fund funded by a severance tax levied on oil and gas produced within the state.[149] The purpose of the fund is to study, protect, and preserve the state's coastal areas.[150] Through this fund a public benefit is clearly realized from oil and gas drilling and directed toward public trust land.

c. *California.* The theory that oil and gas exploration necessarily benefits the public and justifies drilling on lands held in public trust was articulated by the California Supreme Court in *Boone v. Kingsbury.*[151] Before this decision, California had enacted a statute that provided for the granting of permits for mineral exploration upon tidal and submerged lands. Suit was brought in the 1920s by permit applicants after the California State Surveyor General refused to grant any permits, claiming that the statute was beyond the power of the legislature. The surveyor general maintained that the tidelands and submerged lands of California were public trust lands, and that as such they could not be leased to private persons, nor could they be used for any purpose other than a "strictly public use."[152] Furthermore, the surveyor general maintained that leasing the land for oil drilling would interfere with the public's right to unobstructed navigation and fishing over the land and would ultimately result in pollution of the waters and the destruction of fish.[153] The court held the statute to be valid and ordered the surveyor general to issue the permits.[154]

147. La. Rev. Stat. Ann. sec. 30:124 (West 1938).
148. La. Rev. Stat. Ann. sec. 30:172 (West 1928).
149. La. Rev. Stat. Ann. sec. 30:313.1 (West 1987).
150. La. Rev. Stat. Ann. sec. 30.316 (West 1983).
151. 206 Cal. 148, 273 P. 797 (1928), *cert. den'd,* 280 U.S. 517 (1929).
152. Id. at 168, 273 P. at 806.
153. Id.
154. Id. at 194, 273 P. at 817.

In reaching this decision, the court agreed with the legislature that the extraction and production of oil and gas was a "public benefit" and an "aid and furtherance of commerce and navigation,"[155] and therefore a proper use of public trust land. The court made clear that public trust lands could be used for purposes other than navigation, commerce, and fishing:

> Upon an examination of the numerous decisions of the American courts holding that lands covered by navigable waters are held in aid of and for the promotion of commerce, navigation, and fishing, it will be found that superadded to the uses of commerce, navigation, and fishing, when the uses in the case at hand to which the lands are to be devoted are not strictly within the narrow definition of any or all of said terms, said decisions employ such phrases as uses for the "public welfare," "parcels of land as are used in promoting the interests of the public," suggesting that when great public interest may be subserved by the alienation of parcels of tide and submerged lands the state's determination of the question in favor of alienation will not be disturbed by the courts.[156]

The court also rejected the surveyor general's contention that oil and gas drilling would necessarily impede navigation and injure fish and disagreed that it would necessarily interfere with the "paramount public interest in the land and water."[157]

In 1976, many years after *Boone v. Kingsbury,* California enacted the California Coastal Act, which expressly recognizes the importance of gas and oil production and the need to extract these resources from public trust lands.[158] Sections 30260 and 30262 of the act prescribe the conditions under which oil and gas development projects may be operated within the coastal zone. Section 30260 permits oil and gas production even when production might be in conflict with other sections of the act (*i.e.,* those concerned with environmental protection), if "to do otherwise would adversely affect the public welfare." The California Coastal Act has, in effect, codified the holding of *Boone v. Kingsbury.* When considering the use to which a parcel of public trust land should be put, the state may give priority to uses that benefit the public other than those traditionally within the public trust doctrine.[159]

A recent lower court case has reaffirmed *Boone v. Kingsbury* and the broad view of the public trust doctrine incorporated into the California Coastal Act. In *Carstens v. California Coastal Comm'n,*[160] the commission granted a coastal development permit for the construction of a nuclear generating plant. This action

155. Id. at 181, 273 P. at 812.
156. Id. at 192, 273 P. at 816.
157. Id. at 183, 273 P. at 812–13.
158. Cal. Pub. Res. Code sec. 3001.2 (West 1976).
159. See ch. 2, sec. B.
160. 182 Cal. App. 3d 277, 227 Cal. Rptr. 135 (1986).

was challenged as a violation of California's public trust doctrine,[161] and the California Coastal Commission was alleged to have failed in its duty under the public trust doctrine to protect the right of access to public trust tidelands.[162] The court, however, rejected Carstens's position that the commission was bound under the doctrine to preserve recreational access over other possible uses. Citing *Boone v. Kingsbury*, the court held that the public trust doctrine as codified in California permitted the Coastal Commission to consider "commercial as well as recreational and environmental needs in carrying out the public trust doctrine."[163] The court pointed to the California Coastal Act as evidence that the state legislature recognized "the need to balance competing interests in the effort to preserve the coastal zone."[164] These competing interests were identified as public safety, private property rights, orderly economic developmen, environmental concerns, and public access.[165] Finally, the court concluded that the Coastal Commission had properly issued the permit after considering and balancing the various uses to which the public trust land could be put.[166]

5. Aquaculture

Aquaculture is defined as the propagation, cultivation, and harvesting of aquatic plants and animals.[167] This definition encompasses a wide range of activities, including fish farming/hatcheries and shellfish cultivation. Under an expanding concept of the public trust doctrine, these activities may readily be included. The doctrine has traditionally covered the public's right to harvest fish and shellfish. It is only a small step to broaden the doctrine to encourage activities, such as the cultivation of fish, that promote continued abundance. This is particularly appropriate because courts have almost universally held that title to the fish within state waters is held by the state in trust for the people of the state.[168] In administrating this trust, the legislature may pass laws regulating the taking and use of fish for the purpose of both protection and harvest.[169]

The idea that a state may authorize the use of public trust lands for the

161. Id. at 288, 227 Cal. Rptr. at 142.
162. Id. at 284, 227 Cal. Rptr. at 140.
163. Id. at 289, 227 Cal. Rptr. at 143.
164. Id., 227 Cal Rptr. at 143.
165. Id., 227 Cal. Rptr. at 144.
166. Id. at 280, 227 Cal. Rptr. at 142. See ch. 2, secs. D and G for discussions of the deference which courts traditionally accord challenged administrative agency decisions.
167. See Cal. Fish and Game Code sec. 17.
168. See, e.g., *People v. Monterey Fish Products Co.*, 195 Cal. 548, 234 P. 398 (1925); *People v. Horden Co.*, 215 Cal. 54, 8 P.2d 481 (1932); *People v. Glenn Colusa Irr. Dist.*, 127 Cal. App. 30, 15 P.2d 549 (1932); *California Oregon Power Co. v. Superior Court*, 45 Cal.2d 858 (1955), 219 P.2d 455, cert. den'd, 352 U.S. 823 (1956).
169. *Monterey Fish*, 195 Cal. at 563, 234 P. at 404.

propagation and cultivation of fish is not new. In the early 1900s the Supreme Court of Florida held that, while public trust lands may not be sold outright to individuals, the state may grant to an individual limited rights to use public trust lands when such use will encourage new industries and the development of natural and artificial resources.[170] The court conisdered "planting and propagating oysters or shell-fish" an activity that enhanced and improved the rights and interests of the whole people."[171] Therefore, lands held within the public trust could be made available to a private person for undertaking these activities.[172] However, the court was careful to point out that the private use of state public trust land was allowable only so long as it did not unreasonably interfere with the public's use of the area for those purposes preserved by the public trust doctrine.[173] This Florida decision provides an early example of a state using the public trust doctrine to support a state-regulated but privately operated aquacultural effort. Public trust principles both rationalized the private use (in the interest of public welfare) and effectively limited it (to protect other public trust uses).

The idea that public trust lands may be privately used as long as the use promotes the public welfare was taken a step further in a more recent case from Oregon. In *Brusco Towboat Co. v. State of Oregon,*[174] the Oregon State Land Board's practice of leasing public trust lands to private entities for specific uses (including aquaculture) was challenged. The court found that, under the Oregon Constitution, the state was required to manage public trust lands "with the object of obtaining the greatest benefit for the people."[175] The court held that the state could reasonably conclude that, although leasing the land denied to the general public the right to engage in the specific use conducted by the lessee on the leased public trust land, the public benefited overall from the compensation the state earns under the lease.[176] However, like the earlier Florida court, the Oregon court made it clear that the public uses guaranteed under the public trust doctrine were paramount to the leased activity, and therefore the state was limited to entering into leases that would not unreasonably interfere with the public's enjoyment of public trust uses.[177]

The Supreme Judicial Court of Maine has similarly recognized the ability of the state to lease lands under the public trust for private aquaculture operations.

170. *State ex rel. Ellis v. Gerbing,* 56 Fla. 603, 47 So. 353 (1908).
171. Id. at 611, 47 So. at 356.
172. Id. at 612, 47 So. at 356.
173. Id.
174. 30 Or. App. 509, 567 P.2d 1037 (1977).
175. Id. at 519, 567 P.2d at 1044 [quoting Or. Const. art. 8, sec. 5(2)].
176. Id. at 521, 567 P.2d at 1045.
177. Id. at 519, 567 P.2d at 1044.

In *Harding v. Commissioner of Marine Resources,*[178] the court upheld the validity of a lease of public trust land for an aquaculture project, concluding that in granting such a lease the state need only make sure that the lease will not interfere with "other uses of the area," such as fishing, lobstering, shellfishing, swimming, and boating.[179] Since the Maine law requires that public trust lands be managed for the benefit of the public,[180] by upholding the lease, the court essentially held that aquaculture serves the public interest under the public trust.

North Carolina provides a specific statutory exception for shellfishing from its general rule that public trust lands are available to all for navigation, fishing, and commerce. Since the late 1800s the state has issued licenses for shellfish cultivation on submerged public trust lands where no natural beds existed.[181] Under current North Carolina law, "[l]eases are granted in order to increase the use of suitable areas underlying coastal fishing waters for the production of shellfish."[182] The grant of the lease is contingent on a determination that the public will benefit from the lease and subject to the state's duty to conserve marine resources.[183] Public trust lands containing natural shellfish beds are not available for lease,[184] but are to remain open to the public. Thus North Carolina also allows private aquaculture operations to be conducted on public trust lands provided that the public interest is primarily served and other public trust uses are not significantly affected by such operations.

The issue of state leasing for aquaculture purposes becomes more difficult when the land under consideration is privately owned although subject to public trust uses or conditions. In such a situation, the state must assert that aquaculture is an activity that, like navigation, fowling, and fishing, is preserved for the public under the public trust doctrine, notwithstanding conveyances of the area to private ownership. In addition, the state must assert that it can regulate these protected activities.

In a recent Massachusetts case, *Town of Wellfleet v. Glaze,*[185] the town issued licenses for the planting, cultivating, and harvesting of shellfish within a specified area pursuant to state statutory authorization. One of the issues in the case was whether the legislature could authorize towns to issue shellfishing licenses for areas of land privately owned but subject to the public trust. The court found that under the public trust doctrine the public was allowed to fish on private

178. 510 A.2d 533 (Me. 1986).
179. Id. at 536.
180. Id. at 537.
181. *State of North Carolina v. Credle,* slip opinion #480PA87-Hyde.
182. Id. at 11, quoting N.C.G.S. sec. 113-202(a) (1987).
183. N.C.G.S. sec. 113-202(h) (1987).
184. N.C.G.S. sec. 113-202(a) (2) 1987.
185. 406 Mass. 79, 525 N.E.2d 1298 (1988).

lands subject to the trust and that fishing included the right to shellfish.[186] The court also stated that the legislature may enact reasonable regulations restricting the public right to fish, including granting exclusive fishing rights to particular individuals.[187] On these two grounds, the court held that a town could, pursuant to state authorization, issue licenses for shellfishing on privately owned public trust lands.[188] The court qualified its holding in two respects. First, it expressly declined to comment on whether a license to conduct aquaculture, as opposed to shellfishing, on privately owned land would be permissible.[189] Second, the court stated that the license to shellfish must not impair the private rights of the landowner, holding that the landowner had the right to moor boats in the area covered by the license, even if the exercise of this right meant that at low tide the boats would rest on and damage the shellfish beds.[190]

The judicial attitude represented by the *Glaze* decision is a mixed blessing for coastal zone managers desiring to authorize and regulate aquaculture in privately owned areas subject to the public trust. While the case considers shellfishing to be an activity within the purview of the public trust, it is not clear that the court would consider the broader practice of aquaculture to be a protected public trust use. In addition, although the court recognized the right of the state to impose reasonable regulations such as licenses on public trust uses, the court provided a strong indication that when the licensed use conflicts with the private use, the private use (at least if water-dependent) will be favored.

In a recent South Carolina case, the court struck down a permit issued by the South Carolina Coastal Council to the private owner of 660 acres of public trust land to impound the land and thereby block several navigable waterways. The permit was conditioned on the landowner allowing state and federal agencies to perform mariculture experiments on the impounded land.[191] South Carolina challenged the permit, claiming that the impoundment would prevent the public from navigating through the land, a protected public trust use, and that there was no overriding public interest that mitigated this public loss. Although the court expressly recognized the potential public benefit that an aquaculture industry could provide to the state, the court found that the main reason for the impoundment of the land was to allow the private owner to lease additional duck blinds. Therefore, the court concluded that there was no overriding public interest that would justify the loss to the public of the right to navigate through the

---

186. Id. at 84, 525 N.E.2d at 1301.
187. Id.
188. Id.
189. Id. at 84 n.8, 525 N.E.2d 1301 n.8.
190. Id. at 86, 525 N.E.2d at 1301.
191. *State v. South Carolina Coastal Council,* 289 S.C. 445, 346 S.E.2d 716 (1986).

public trust land, and accordingly it held that the council lacked the authority to issue the permit.[192] Although the court thus appeared to prevent state coastal resource managers from using public trust land for aquaculture purposes, the court in fact made it clear that it doubted whether aquaculture was the motivating factor behind this particular permitting arrangement. Moreover, the court's opinion indicated that it viewed aquaculture to be a potential source of great public benefit, and that if aquaculture were the bona fide use to which public trust land would be put, such use might be permitted even if another public trust use such as navigation was lost.

California, through the California Coastal Act, has enacted a statutory framework, grounded in public trust principles, for the promotion and regulation of aquaculture industries. California Public Resources Code section 30001.5 (West 1986) provides the several goals of the act, one of which is to "[p]rotect, maintain, and, where feasible, enhance and restore . . . natural and artificial resources." The cultivation of marine life is an activity that would further this stated goal. Section 30230 of the act pertaining to marine resources reiterates the state's concern for the maintenance and continuation of marine life, providing that the marine environment shall be used in a way "that will maintain healthy populations of all species of marine organisms." Aquaculture is specifically provided for in the act under section 30222.5, which states that "[o]cean front land that is suitable for coastal dependent aquaculture shall be protected for that use, and proposals for aquaculture facilities located on those sites shall be given priority, except over other coastal dependent developments or uses." Hence, this statute provides coastal zone managers with an express statement by the legislature that aquaculture is a priority use on suitable public trust lands. Further legislative support for aquaculture is provided in the California Fish and Game Code, division 12 of which has, since 1982, been devoted exclusively to aquaculture. Under California Fish and Game Code section 15100 (West Supp. 1990), the Aquaculture Development Section is created within the California Department of Fish and Game, to regulate and provide information on the aquaculture industry. California Fish and Game Code section 15400 (West Supp. 1990) provides for the leasing of state water bottoms for aquaculture purposes.

The concerns reflected in these California provisions are similar to those expressed in the North Carolina statutes previously discussed. State water bottoms may be leased for aquaculture purposes provided that the lease is determined to be in the public interest. Further restraint on the state's ability to lease public trust land for aquaculture is provided by section 15401, which requires the state to designate those areas that have been used by the public for digging

192. *South Carolina Coastal Council,* 289 S.C. at 451, 346 S.E.2d at 720.

clams, and forbids the state from leasing these areas for aquaculture. Section 15402 gives the lessee the "exclusive right to cultivate and harvest the aquatic organisms in the area leased."

Consistent with traditional public trust principles, an initial determination must be made that the public interest is furthered by the lease. Further, lessees may not unreasonably interfere with public use of the area for fishing, navigation, commerce, or recreation (under section 15411 of the act). However, where uses conflict the statute appears to tilt toward the private use: "The lessee may . . . limit public access to the extent necessary to avoid damage to the leasehold and the aquatic life culture therein."[193] Thus California has, in the case of aquaculture, provided an example of utilizing public trust principles to benefit the public by favoring private activities.

6. Environmental Protection and Estuaries

Environmental protection can be said to fit well within the scope of the public trust doctrine. Federal and state courts have acknowledged that public trust interests and uses protected by the doctrine now include preservation of public trust lands and waters in their natural state,[194] of their ecology,[195] and of their indigenous flora and fauna.[196] The courts have also noted that the doctrine has "evolved from a primarily negative restraint on states' ability to alienate trust lands into a source of positive state duties."[197] The scope of public trust authority to protect environmental resources is potentially very broad. Some state courts, for example, have upheld regulation or even prohibition of certain activities in order to protect water quality.[198] Such regulatory authority may reach beyond the main water body to its tributaries,[199] and arguably to its shores where land uses

193. California Fish and Game Code sec. 15400 (West Supp. 1990).

194. *District of Columbia v. Air Florida, Inc.,* 750 F.2d 1077, 1083 (D.C. Cir. 1984); *State v. Superior Court of Lake County,* 625 P.2d 239, 250–51, 29 Cal.3d 210, 229–30 (1981); *Marks v. Whitney,* 6 Cal.3d 251, 259–60, 98 Cal. Rptr. 790, 796, 491 P.2d 374, 380 (1971).

195. *City of Berkeley,* 26 Cal.3d at 521, 162 Cal. Rptr. at 330–31, 606 P.2d at 364–65; *Wisconsin Environmental Decade, Inc. v. Department of Natural Resources,* 85 Wis.2d 518, 526 (1978), 271 N.W.2d 69, 72; *Marks,* 6 Cal.3d at 259, 98 Cal. Rptr. at 796, 491 P.2d at 380.

196. *Air Florida,* 750 F.2d at 1083.

197. Id.

198. *Mono Lake,* 33 Cal.3d 419, 189 Cal. Rptr. 346, 658 P.2d 709 (diversion of water from tributaries of Mono Lake, lowering the water level and increasing salinity); *Kootenai,* 105 Idaho 622, 671 P.2d 1085 (moratorium on permits for encroachments on lake to protect lake's water quality).

199. *Mono Lake,* 33 Cal.3d 419, 189 Cal. Rptr. 346, 658 P.2d 709; but cf., *Golden Feather Community Ass'n v. Thermalito Irrigation Dist.,* 209 Cal. App. 3d 1276, 257 Cal. Rptr. 836 (1989) (doctrine does not apply where impact of diversion on reservoir is not impact on waters subject to public trust).

impact on water quality through, for example, leaching and runoff, extraction, and waste water disposal.[200]

To date, the courts do not appear to have approved an expansion of the doctrine's scope to include privately owned uplands for the purpose of water quality protection. In the absence of case law on this particular issue, an analogy could be drawn to the issue of public access to the shoreline and efforts to make the public trust rights "meaningful" by guaranteeing perpendicular access over adjoining uplands. Decisions expanding the public's right to that extent, however, are still in the minority and, because of the impact on private property rights, require a case-by-case balancing of interests.[201] Even within a more narrow scope, applying the public trust doctrine to water quality protection raises problems which the courts might have to resolve on a case-by-case basis. While water quality regulations would serve the public interest in one respect, they are likely to restrict other uses that the doctrine protects. Decisions as to which lands and waters are subject to public trust authority and which uses will be restricted are bound to be challenged in court.[202]

Case-by-case judicial review can be avoided or at least minimized if state legislatures enact public trust programs. Such legislation might simply mandate regulation of land and water uses in order to protect a particular environmental interest recognized by the legislature as a public trust interest. Articulating a clear public trust policy can complement a state's existing police power to regulate land and water use in the interest of the public's health, safety, and welfare. Simply declaring a policy, however, does not resolve potential conflicts between different uses of or interests in public trust lands and waters. State agency decisions are still likely to be challenged and the balancing of interests ultimately may still have to be done by the courts, on a case-by-case basis.

A more powerful application of the public trust doctrine to environmental protection can be created if a state legislature authorizes special or critical area management programs within the state. Such legislation can declare environmental quality protection a public trust interest and identify lands and waters of special or critical value to that interest. Areas may be designated special or critical for one specific value, such as drinking water quality, but more typically represent a combination of values. Rather than regulation to protect one endangered species, reduce the level of a certain pollutant, or champion one particular use,

200. Johnson, "Water Quality Control by the Public Trust Doctrine," published in D. Getches, ed., Water and the American West, pp. 127–42 (1988).

201. *Bay Head Improvement,* 95 N.J. 306, 471 A.2d 355; See also ch. 2, secs. A and H, and ch. 3, sec. C, subsection 1.

202. See Johnson at 134.

the goal under public trust legislation should be to preserve an entire area for all of its inherent values.

Although designated areas must lie within the judicially accepted geographic scope of the public trust doctrine, authority to protect the environmental quality can reach further. Public trust legislation not only strengthens a state's authority to regulate uses within management areas, but also strengthens state and local authority to regulate and plan uses of adjacent lands and waters because activities on those lands and waters will impact public trust lands and resources. As part of comprehensive management plans, state and local governments could designate buffer zones around special or critical resource management areas, deriving authority from both the public trust doctrine and their police powers.[203]

This comprehensive management approach to environmental protection is likely to reduce legal challenges and to shift the focus of judicial review. Instead of scrutinizing a state agency's balancing of competing interests in a particular case,[204] a court will defer to the balancing of interests articulated in legislation and a comprehensive plan. As long as the legislature provides sufficient guidelines for the development of a management plan,[205] and the agency's decision is consistent with those guidelines and with the plan, the agency's exercise of authority is likely to be upheld and therefore less likely to be challenged.[206]

While a comprehensive legislative and management scheme may reduce the number of legal challenges to environmental regulation, it does not "subsume" the public trust doctrine. Independent from a public trust-oriented statutory and regulatory program, the common law public trust doctrine operates as a source of authority for the state to intervene in the use of public trust resources, a bar to anyone acquiring a vested right in such resources in a manner that is harmful to the public right, and a "standing" doctrine through which members of the public—the beneficiaries of the trust—may enforce the state's trustee obligations.[207]

203. In other words, the public trust doctrine and legislation grounded in it can enhance a state's ability to exercise its police powers and shield it from constitutional challenges. This new application is an extension of the traditional role of the doctrine to avoid erosion of the sovereign's right and duty to protect the public interest. See generally ch. 3, secs. A and B for area designation and discussion of comprehensive management.

204. E.g., *Billings v. California Coastal Comm'n,* 103 Cal. App. 3d 729, 163 Cal. Rptr. 288 (1980); *Morse v. Oregon Division of State Lands,* 285 Or. 197, 590 P.2d 709 (1978); *Saxon v. Division of State Lands,* 31 Or. App. 511, 570 P.2d 1197 (1977).

205. As a matter of administrative law the legislature may delegate authority to an executive agency only if it provides or retains ultimate supervisory control over the agency's decisions, which it typically does by providing it with adequate guidance for those decisions. See ch. 2, sec. D.

206. See *Adams v. North Carolina Dep't of Natural and Economic Resources,* 295 N.C. 683, 249 S.E.2d 402 (1978); *Candlestick Properties, Inc. v. San Francisco Bay Conservation and Dev. Comm'n,* 11 Cal. App. 3d 557, 89 Cal. Rptr. 897 (1970).

207. *Mono Lake,* 33 Cal.3d 419, 189 Cal. Rptr. 346, 658 P.2d 709. See also Sax, "The Public Trust

In short, a state may find substantial additional authority in the public trust doctrine to advance an environmental protection program. In enacting a clear public trust policy, a state can mandate regulations aimed at specific environmental quality goals. Alternatively, a state legislature can authorize comprehensive management plans to preserve entire areas of "special" or "critical" environmental value. Such legislation must contain specific findings regarding the environmental significance of public trust resources, declare a policy of balancing competing public interests so as to enhance and preserve those resources, and provide clear guidelines for the development and implementation of a management plan. The application of the public trust doctrine to the management of estuaries is an illustration of such an approach.

Estuaries are partially enclosed coastal water bodies where salt water from the ocean mixes with fresh waters from inland. Estuarine areas include river mouths and sea inlets, bays, sounds, harbors, lagoons, tidal marshes, coasts, and inshore waters subject to tidal action. Scientists and environmentalists have come to recognize that estuaries are unique ecosystems of critical value to human activity and to wildlife. They provide shelter from wave action and oceanic predators. Also, because of their shallow depth, estuaries provide stratification and circulation of nutrient-rich waters, as well as storage and slow release of nutrients, forming a vital link in the food-energy cycle. The biological productivity of estuaries is of direct importance to offshore fishing, both commercial and recreational. Besides sustaining a rich variety of plant and animal life, estuaries purify water supplies, absorb pollution, prevent siltation of nearby rivers and streams, and prevent erosion of stream banks.[208]

The estuarine ecosystem includes estuary wetlands shallow enough to invite diking and filling, tideland wetlands between the mean low watermarks and high watermarks, and tidal marsh wetlands ranging from the lowest high watermark to the nonaquatic vegetation line. These boundaries do not always match those that define legal rights under the public trust doctrine. Although ownership rights may vary by state, estuarine ecosystems usually include privately owned lands both subject to and free from public trust restrictions as well as publicly owned lands, with property lines not necessarily following ecological boundaries.

The natural and economic resource value of estuaries is important to a state

---

Doctrine in Natural Resource Law: Effective Judicial Intervention," 68 Mich. L. Rev. 471 (1970); see also ch. 2, sec. E. While state agencies may not wish to emphasize this latter role, the right to raise, on behalf of the public, a claim of harm to the public trust may well be an essential part of the doctrine's system of checks and balances. Expansion of state authority may be acceptable to the courts in part because the doctrine provides such a "check." See *Mono Lake*, 33 Cal.3d at 431 n.11, 189 Cal. Rptr. 353 n.11, 658 P.2d at 716 n.11.

208. See Hildreth and Johnson, *Ocean and Coastal Law*, 7–11 (1982); Chapman, "Regulating Fills in Estuaries: The Public Trust Doctrine in Oregon," 61 Or. L. Rev. 523, 523–24 (1982).

whether the estuaries are just a small portion of a state's coastal zone, as in Oregon, or occupy vast areas, as in Louisiana and North Carolina. Yet only in recent decades have the states begun to recognize their value as ecosystems. Historically, and still today, protected estuary waters have been valued as ideal sites for harbors and have attracted development associated with both commercial and recreational navigation. Development led to diking and filling, reducing overall wetland area, interfering with the quality and productivity of the remaining lands and waters, and frequently destroying ecological integrity.[209] Similarly, dredging to maintain harbors and navigation channels not only affects the quality and productivity of estuarine waters and shellfish beds, but also contributes to the filling of estuaries and to subsequent disputes regarding ownership rights.[210]

Increasingly, such disputes will be resolved under both the common law public trust doctrine and public trust oriented statutes. In this section we suggest that: the public trust doctrine provides an appropriate mechanism to balance conflicting interests in estuarine resources; the balancing can be done most effectively within the context of comprehensive special or critical area management plans mandated by state legislatures explicitly exercising their public trust authority; and the common law public trust doctrine should be used independently, less as a means to resolve individual disputes and more as a check on legislative delegation of authority or on the exercise of such authority by management agencies.

There can be no doubt that recent court decisions have brought preservation of estuarine ecosystems within the scope of the common law public trust doctrine. In addition, the value of estuaries has been recognized explicitly in federal and state statutes. The National Estuary Program (NEP), first proposed in 1985 by EPA as part of its Near Coastal Water Initiative, was incorporated into the Federal Water Pollution Control Act (FWPCA) in 1987.[211]

In 1986 an amendment to the Coastal Zone Management Act revamped the

209. Between 1850 and 1960 the open surface water area of San Francisco Bay was reduced from 680 to 430 square miles at an average rate of 3½ square miles a year. Hildreth and Johnson, note 12, at 470–71. From 1930 to 1971 fills reduced Oregon's tidelands by approximately 16 percent. Chapman, 61 Or. L. Rev. at 524 n.9. See also North Carolina Coastal Area Management Act, N.C.G.S. sec. 113A–102 (1983), quoted in Hildreth and Johnson at 433.

210. E.g., *City of Alameda v. Todd Shipyards Corp.*, 632 F. Supp. 333 (N.D. Cal. 1986). See also ch. 3, sec. C, subsection 7.

211. Chapman, 61 Or. L. Rev. at 523 n.1. Under this program, states may nominate estuaries of "national significance" and request management conferences to develop a comprehensive management plan. If the request is granted, the state is eligible for financial assistance. The program focuses on controlling point and nonpoint sources of pollution to protect public water supplies as well as indigenous shellfish, fish, and wildlife populations. As of November 1988, sixteen national estuaries have been designated: Long Island Sound, Narragansett Bay, Buzzards Bay, Massachusetts Bay, Puget Sound, New York/New Jersey Harbor, Delaware Bay, Delaware Inland Bays, Albemarle Sound, Sarasota Bay, San Francisco Bay, Santa Monica Bay, Galveston Bay, Barataria-Terrebonne Bay, Indian River Lagoon, and Peconic Bay. 33 U.S.C. sec. 1330(a)(2) (B) (Supp. 1989).

existing Estuarine Sanctuary Program (since 1972) and established the National Estuarine Reserve Research System under which states may nominate "representative" estuarine ecosystems and apply for land acquisition and management grants.[212] As does the NEP, this program focuses mainly on research and educational projects, although limited funds may also be used by states to acquire property and operate reserves.[213]

While the states have primary responsibility for the management of their estuaries, the FWPCA and CZMA express a national public interest in their protection.[214] For example, because estuarine water quality is critical to certain fisheries of national importance, federal programs to coordinate cleanup efforts have been established for Chesapeake Bay and the Great Lakes.[215] Thus, while both statutes mandate state management programs,[216] states carry out their management responsibilities subject to a national public interest.[217]

In response to federal requirements and to ad hoc judicial decision making, several states have attempted to define their public trust authority more clearly. Some states have explicitly extinguished their public trust interest in previously filled tidelands that have been conveyed to private persons and are "landlocked"[218] or narrowed their authority to restrict non-water-related uses on such lands.[219] Many other states have passed laws asserting a continuing public trust interest in coastal resources, including estuaries, which are still available or adaptable for public trust uses.[220] Typically, these laws incorporate a "public

212. 16 U.S.C. sec. 1461 (Amend. Supp. 1989).

213. 16 U.S.C. sec. 146(e) (Amend. Supp. 1989).

214. 33 U.S.C. sec. 1330(a) (2) (A) (Amend. Supp. 1989); 16 U.S.C. sec. 1451(a) (1980).

215. 33 U.S.C. secs. 1267, 1268 (Amend. Supp. 1989).

216. 16 U.S.C. secs. 1451(i), 1452(2) (1980); 33 U.S.C. secs. 1085, 1288, 1330(a) (1) (Amend. Supp. 1989).

217. See *United States v. 1.58 Acres*, 523 F. Supp. 120 (D. Mass. 1981) (state's administration of public trust and land below low watermark is subject to paramount rights of federal government with respect to matters within its powers). Note that the FWPCA and the CSMA clearly place protection of natural resources within the federal powers, while the Submerged Lands Act of 1953 explicitly left "the rights of management, administration, leasing, use, and development" of submerged lands and natural resources within state territories to the states. 43 U.S.C. secs. 1311, 1314 (1953).

218. Mass. Gen. L. ch. 91 (Amend. 1983), partly in response to *Boston Waterfront Development Corp. v. Massachusetts*, 378 Mass. 629, 393 N.E.2d 356 (1979); see Lahey, "Waterfront Development and the Public Trust Doctrine," 70 Mass. L. Rev. 55 (1985). But see *Opinion of the Justices*, 437 A.2d 597 (Me. 1981) (holding that amended Me. Rev. Stat. Ann. tit. 12, sec. 559, releasing state public trust rights in lands filled prior to October 1, 1975, would not violate the Maine constitution or the state's public trust duties).

219. Oregon Fill and Removal Act, Or. Rev. Stat. sec. 541.605.691 (1979) amendments in response to *Morse v. Oregon Division of State Lands*, 285 Or. 197, 590 P.2d 709 (1978), see Chapman, 61 Or. L. Rev., at 542–44.

220. E.g., Washington Shoreline Management Act, Wash. Rev. Code sec. 90.58 (1971) (preservation of "shoreline of statewide significance"); California Coastal Act, Cal. Pub. Res. Code sec. 27001

purpose" test developed by the courts under the common law public trust doctrine. Under such a test the proposed use must not unreasonably interfere with the statutory policy, i.e., must be either water-dependent use or have minimal impact on public trust lands and resources; or the use must be for a public purpose and the public's need for non-water-dependent use must outweigh impairment of statutorily defined public trust uses.[221]

This expression of state public trust authority often takes place in the context of, or requires consistency with, the state coastal zone management program (CMP).[222] The significance of this is twofold. First, approval of a CMP under the federal CZMA strengthens a state's authority to plan and regulate uses of coastal resources without being preempted by other federal statutes.[223] Second, the federal CZMA mandates comprehensive management plans[224] and encourages preparation of special area management plans.[225] Following this mandate, several states have designated estuaries as "areas of environmental concern" or "critical areas," or otherwise developed management plans to preserve the quality and integrity of entire estuary ecosystems.[226]

It is clear that comprehensive management plans enhance a state's authority to protect and manage its estuaries. If they are consistent with an approved coastal management plan, planning and regulatory decisions may be shielded from taking and preemption claims. If developed as a comprehensive scheme of state and local control over land and water uses, and if mandated and guided by the legislature invoking public trust principles, the plan's implementation will survive a claim of improper delegation of authority and avoid strict scrutiny by the courts.[227] In 1970 the California Court of Appeals upheld a denial by the San

---

(1972) (permanent protection of coastal zones natural resources and ecosystem of "paramount concern" to present and future residents); Oregon Fill and Removal Act, Or. Rev. Stat. sec. 541.605.695 (Amend. 1981) (paramount policy to preserve the use of its waters for navigation, fishing and public recreation); Mass. Gen. L. ch. 91, sec. 2 (Amend. Supp. 1989) ("[p]reserve and protect the rights in tidelands of the inhabitants of the Commonwealth by insuring that the tidelands are utilized for water dependent uses or otherwise serve a proper public purpose"); see also Mass. Gen. L. ch. 131, sec. 40 (Amend. Supp. 1989) (Wetlands Protection Act) and Mass. Regs. Code tit. 310, secs. 10.21–10.37 (1989) (Coastal Wetlands Regulations).

221. See ch. 5.

222. See Mass. Regs. Code tit. 310, sec. 10.22 (1989).

223. See *Friends of the Earth v. U.S. Navy,* 841 F.2d 927 (9th Cir. 1988) (upholding application of state dredging regulations, as well as FWPCA regulations, to Navy activities on federal lands in coastal zones).

224. 16 U.S.C. secs. 1451(i), 1452(2) (Amend. Supp. 1989).

225. 16 U.S.C. sec. 1552(3) (Amend. Supp. 1989).

226. E.g., N.C.G.S. secs. 113A-113-115 (1987); Chesapeake Bay Critical Area Protection Program, Md. Nat. Res. Code sec. 1813 (1983); Or. Rev. Stat. secs. 197.405 (1989); Wash. Admin. Code sec. 173-16-040; Mass. Regs. Code tit. 310, secs. 10.23–10.24 (1989).

227. See ch. 2, sec. D.

Francisco Bay Conservation and Development Commission (BCDC) of a fill permit. The BCDC had developed a comprehensive bay plan, which was adopted by the legislature in 1969 (and later incorporated into the state's coastal management program). The plan spelled out general policies to preserve the bay as a great natural resource for present and future generations and to develop the bay and its shorelines to their highest potential with a minimum of filling, as well as specific planning goals and regulatory guidelines. The plan also provided maps indicating appropriate uses for different parts of the bay area.[228] The court held that legislation creating the BCDC, as well as denial of the permit consistent with legislative policy to develop a conservation and development plan for the bay, was a valid exercise of the police power.[229]

In 1987 the Supreme Court of Washington considered a challenge to comprehensive plans of the state and Skagit County which precluded dredging and filling in tidelands owned by the plaintiff.[230] The property was located in Padilla Bay, an estuary on Puget Sound, included in the "shorelines of statewide significance" under the 1971 Shoreline Management Act. In 1976 Padilla Bay was identified as an area of particular concern under the Coastal Zone Management Program. Dredging and filling were determined inappropriate uses for "aquatic" property. Finally, in 1980 the legislature established the Padilla Bay Estuary Sanctuary to preserve existing tidelands, including plaintiffs' property. The issue before the court was whether application of the provisions in the state and county management plans and the sanctuary program constituted a regulatory taking.[231] The court held that plaintiff had acquired the property subject to the public trust that existed prior to any of the legislative schemes, but remanded for a factual determination of the restrictions imposed by the public trust on the owner's use of the property. If additional facts showed that plaintiff "could make no reasonably profitable use of its land consistent with the public trust, no taking occurred."[232] The court's analysis clearly suggests that the public trust doctrine can be used to identify uses that injure the public interest and to insulate regulations aimed at protecting that interest from taking challenges.

Also in 1987 the Fourth Circuit Court of Appeals upheld the district court's abstention from deciding a challenge to a county's refusal to permit residential development on a portion of the developer's land pursuant to the Chesapeake Bay Critical Area Protection Program.[233] Abstention was upheld on two grounds.

228. Hildreth and Johnson at 475–77.
229. *Candlestick Properties*, 11 Cal. App. 3d 557, 89 Cal. Rptr. 897, 905 (1970).
230. *Orion*, 109 Wash.2d 621, 747 P.2d 1062.
231. For discussion of takings issue, see ch. 2, sec. H.
232. Id. at 660, 747 P.2d at 1083.
233. *Meredith v. Talbot County*, 828 F.2d 228 (4th Cir. 1987).

First, federal courts should not interfere with a comprehensive state regulatory scheme concerning state and local land use policies. Second, the federal constitutional taking claim could be avoided by interpretation of state laws contained in the Chesapeake Bay Critical Area Protection Program, which the state courts had not yet considered.[234] Finally, a public trust-oriented management plan may prove to be a flexible tool, capable of adapting to evolving perceptions of public trust interests, management priorities, and management principles.[235]

In addition to providing insulation from judicial challenge, the public trust doctrine provides a balanced, comprehensive approach to estuary management. A balancing of various public trust interests, as well as other public purposes, is inherent in the doctrine. Public trust-oriented management should seek to accommodate a variety of interests, federal, state, and local, private and public, consistent with a clearly established state policy.

For an estuary management program to maximize the benefits from the public trust approach, it is important that:

1. the legislature clearly expresses its public trust interest; this requires (a) specific legislative finding regarding the unique characteristics and values of estuarine resources, and (b) a clear declaration of legislative policies and goals, for example, to protect these resources for present and future generations by encouraging appropriate uses or to safeguard the ecological integrity of estuaries;
2. the legislature mandates comprehensive plans for the coastal area which identify areas of special or critical concern such as estuaries, appropriate as well as inappropriate uses, and local versus state jurisdiction; and
3. the legislation and the management plan, including the special area management plans, provide for areawide balancing of public trust uses and other uses that benefit the public, and for substantial public review thereof.

7. Harbor Development and Maintenance

Coastal zone management consists in great part of resolving actual and potential conflicts between private and public rights; local, state, and federal authority;

234. Id. at 231–32. The applicants had alleged violations of state and local law and asked for a mandatory injunction requiring the county to approve the development; the federal constitutional taking issue would disappear if a state court granted the injunction after interpreting a newly enacted section of the bay program. The holding, stated as an application of general principles, suggests that federal courts will defer to state legislative and judicial powers to enact a comprehensive management scheme and to resolve conflicts arising under such a scheme.

235. See *City of Berkeley,* 26 Cal.3d 515, 162 Cal. Rptr. 327, 606 P.2d 362. A public trust legal foundation for natural resources management could provide much support for "adaptive management" as well as management based on ongoing "carrying capacity" analysis.

alternative uses of resources; and the present versus the future quality and availability of those resources. The development and maintenance of harbors and marinas, particularly waterfronts in urban areas, raise all these issues with a particular degree of difficulty. State coastal managers face a seemingly monumental challenge to balance and reconcile well-established, often virtually exclusive, private use of ostensibly public resources with public interest objectives such as environmental quality, preservation of ecologically significant resources, and protection of traditional public trust activities, such as local fisheries. Their exercise of authority is additionally complicated by dominant federal power over navigable waters.

Harbors are of vital importance to local and interstate commerce. To serve their function, they must provide both shelter and access. While natural harbors may offer both without much alteration of coastal lands or waters, the establishment of harbors more commonly requires dredging and filling of estuaries or other tidelands. Such alterations are likely to have a physical as well as ecological impact, not only in their specific location but in a larger coastal area, including impacts on water flow, sediment movement, rates of erosion or accretion, wildlife habitat, and access to and quality of fishing grounds. Subsequent use and development of harbors, even of natural ones, has a continuing affect on nearby coastal resources, for example through gradual environmental pollution and through continuing conversion of land and water to residential and recreational as well as commercial uses. Sediment at the harbor bottom accumulates pollutants to a degree that some harbors have been placed on priority lists for cleanup of hazardous waste. Where maintenance of navigational channels in the harbor area stirs up such sediments, the short- and long-term effects on water quality can be serious. Unless carefully planned and regulated, the development and maintenance of harbors can degrade the value of a coastal area as a natural and public resource and severely restrict land and water use in the harbor area.

Similarly, marinas, an omnipresent feature of most harbors, serve an ever-increasing demand for recreational boating facilities and are therefore basically consistent with the public interest in facilitating access to navigational opportunities. However, their construction and maintenance can have a negative impact on the environmental quality and resource value of the surrounding area, adversely affecting navigation, fisheries, and the preservation of environmentally sensitive areas. Moreover, their exclusive occupation of limited shorefront space more often than not seems to lead to restricting rather than expanding public access to the shore. Examples are residential marina communities and the concept of dockominiums or boat slips available for exclusive private ownership.[236]

236. See ch. 3, sec. C, subsection 3.

This trend provides a vivid illustration of the dilemmas and the extraordinarily difficult balancing process facing those charged with managing public trust lands and resources. Approval of short-term uses may mean the degradation of public trust lands and resources, such that they are not available for use in the future;[237] use for one valid purpose can exclude other uses or users.[238] This aspect of harbor and marina development is the most difficult to manage. It no longer suffices to attribute negative impacts to discrete events taking place at one point in time. The potential negative impact from marinas may be less a function of infrequent dredging or discharges from individual vessels and facilities than of rapid and unbalanced development of a coastal area which ignores or exceeds that area's carrying capacity. Similarly, economic development in harbor areas can reach a point where the cumulative demand on natural resources exceeds and degrades their capacity.

Many state statutes grant local governments the authority to develop their harbor areas and to use, sell, or lease coastal lands or waters for that specific purpose,[239] or more generally for a public use or purpose.[240] State constitutions or statutes may also provide for the establishment of harbor lines, which affect both local authority and private ownership rights. Historically, bulkhead and pierhead lines may have marked the seaward limits where riparian owners could fill or build open-pile structures without a permit. Modern schemes usually draw inner and outer harbor lines that define the harbor bed.[241] Beyond the outer harbor, the state or local government may not permit any structure, dredging, or filling.[242] Submerged lands between the inner and outer harbor lines may be reserved by the state for landings, wharfs, streets, or other uses related to commerce and navigation, restricting both public and private activity in the harbor area.[243] The inner harbor line usually excludes privately held property.[244] Other states may have more simple arrangements, drawing harbor lines beyond which pier, wharf, or other structures may not be extended without a state license.[245]

237. See ch. 2, sec. H for discussion of present versus future uses of public trust lands and resources.

238. A related complication for planners and managers is that particular interests of groups of users often cannot be categorized conveniently as associated with either a positive or a negative impact. For example, the interests of riparian landowners may conflict with conservation or public access policies in one case; in another it may be necessary to protect riparian rights in the face of potential overuse, by recreational or commercial navigators, of a limited resource capacity.

239. E.g., *California ex rel. State Lands Comm'n v. County of Orange*, 134 Cal. App. 3d 20 (1982).

240. E.g., *Port Chester Yacht Club v. Port Chester*, 507 N.Y.S.2d 465 (N.Y. App. A. D. 1986).

241. See generally, Johnson and Cooney, "Harbor Lines and the Public Trust Doctrine in Washington Navigable Waters," 54 Wash. L. Rev. 275 (1979).

242. At present, state harbor lines will usually coincide with harbor lines established by the U.S. Army Corps, beyond which a federal permit is required for construction, dredging, and filling. Id.

243. E.g., Wash. Const. art. 15, secs. 1–3.

244. Johnson and Cooney, at 290.

245. E.g., Mass. Gen. L. ch. 91, sec. 34.

Restrictions on land and water use in any harbor line system, however, are typically based on the basic fact of public ownership rather than on the more expansive concept of public trusteeship. A harborline system does not, therefore, by itself define the rights of landowners over waters adjoining their properties landward of the inner harborline. Instead, these rights are determined by the terms of statutory conveyances out of, or subject to, the public trust and by the common law public trust doctrine.[246] Similarly, a harbor line system does not provide a mechanism to rank priorities or balance competing uses in harbor areas consistent with navigation, commerce, or other public policies. That mechanism may, however, be derived from the public trust doctrine. One of the most effective methods is to establish criteria, by statute or agency regulation, stressing the priority of water-dependent uses, a particularly valuable technique in managing waterfront development in urban harbor areas.

For decades, urban harbors and waterfronts have seen the economic and physical decline and decay of traditional water-dependent activities favored by the public trust doctrine, such as commercial fishing, fish processing, and boat servicing and repairing. Waterfronts became the least attractive, cheapest areas in many communities and consequently were increasingly marked by deteriorating wharves and warehouses, vacant buildings, and incinerators and power plants deemed undesirable elsewhere. With the growth of the urban development and the environmental and historic preservation movements of the 1960s and 1970s, however, waterfronts underwent a rebirth as desirable locations. Unfortunately, those historical activities that inherently required waterfront locations typically have been unable to compete with the non-water-dependent uses such as high-density housing, offices, retail stores, hotels, and other commercial activities that have proliferated as parts of waterfront renewal projects—activities that do not need a waterfront location but seek to maximize their value by taking advantage of water vistas and maritime ambience. In addition, the employment, tax, and other economic benefits to a community of such waterfront developments (not to mention the common political influence of the sponsors of such projects) can easily overshadow the concurrent threats they pose to fauna and flora that require coastal habitat.

In reaching difficult decisions about how to allocate the limited waterfront resources among competing harbor uses (within the guidelines and restrictions

---

246. To what extent a management program may restrict riparian rights depends on the purpose of the restriction, to what extent the restriction is within the scope of the purpose, and possibly to what extent the owner has exercised rights previously to make improvements on the property. Restricting an activity that interferes with navigation or commerce or with public safety and health does not amount to a taking, whereas an invasion of private property to acquire access to navigable waters may require just compensation. See, e.g., *Pacific Legal Foundation v. California Coastal Comm'n*, 128 Cal. App. 3d 694 (1982). See discussion in ch. 3, sec. C, subsection 3.

provided by statutes and zoning codes), coastal managers can make effective use of the public trust doctrine, because it affords them a legal basis for preferring water-dependent uses and preventing undue encroachment of non-water-dependent and private development near the water's edge without running afoul of takings challenges from private property owners.[247] In particular, the public trust doctrine allows them to discourage uses inconsistent with the public trust even where the state has conveyed public trust land on the waterfront to private parties;[248] and even where federal officials have sanctioned such uses without proper consideration of their consistency with public trust principles contained in the states coastal zone management plan.[249]

247. See ch. 2, sec. H.

248. See, e.g., *Orion,* 693 P.2d 1369 (Wash. 1987), *cert. den'd,* 108 S.Ct. 1996 (1988); *Boston Waterfront,* 378 Mass. 629 (1979).

249. See generally the discussions in ch. 2, secs. D and G, ch. 4, sec. A, subsection 5, and ch. 4, sec. B. For discussion of a state program emphasizing the criterion of water dependency, see ch. 5.

IV

The Public Trust Doctrine and Federal-State Relations

The public trust doctrine, like any other state law, operates within our constitutional framework of government. Use of the doctrine as a tool of coastal management inevitably leads to situations in which state and federal authorities are, or appear to be, in conflict. How might this conflict be resolved? Can states rely upon public trust principles to modify or prohibit activities conducted or permitted by federal agencies that adversely affect trust lands and resources? The federal government is itself the title owner of large areas of coastal lands originally burdened by the public trust. Does the doctrine impose any duties on the federal government with respect to these lands?

This chapter addresses these questions. Section A discusses federal-state relations generally in the context of the supremacy of the federal constitution, laws, and treaties over state law, as well as the circumstances in which federal law may preempt state law. Section B analyzes how the states may use their federal consistency authority under the CZMA to assert public trust duties against federal agencies and federally permitted projects. Finally, section C considers the trust obligations of the federal government itself as the holder of title to lands originally subject to the public trust.

A. SUPREMACY AND PREEMPTION

This section addresses problems in managing trust lands, waters, and resources caused by intergovernmental conflicts. While the states have sovereignty over their tidelands and submerged lands, their authority to regulate them is always subject to the supreme authority of the federal government under the United States Constitution. State actions that conflict or are incompatible with federal legislative policy are held to be preempted by federal law and are of no effect. Such preemption has commonly occurred in the areas most relevant to the public trust doctrine, regulation of maritime commerce and navigation, where the powers of Congress are plenary. This section discusses the ways in which federal preemption of state regulatory authority has been or can be avoided or minimized, including reliance on liberal judicial rules of interpretation, the judicial doctrine of federal and state cotrusteeship, and taking advantage of federal waivers of sovereign immunity, so that coastal managers and advocates can support state regulation implementing public trust principles with relatively little concern about running afoul of federal preemption, even as to federally conducted or permitted activities.

1. Generally

According to well-known United States Supreme Court decisions, it is the settled law of the United States that the states own and have dominion and sovereignty over tidelands and submerged lands and over the waters and submerged lands of navigable freshwater rivers, streams, and lakes for the benefit of their citizens.[1] But the Supreme Court has also made it clear that the states exercise their trusteeship responsibilities over these lands subject to the paramount right of Congress concerning certain maritime matters that the states have surrendered to the federal government in the Constitution (most significantly, regulation of interstate commerce and navigation, often termed the "federal navigational servitude"), by virtue of the supremacy clause of that Constitution.[2]

The supremacy clause of the Constitution (art. VI, cl. 2) provides that "[t]his Constitution, and the Laws of the United States which shall be made in pursuance thereof; and all treaties made, . . . shall be the supreme law of the Land." State laws that are incompatible with the policy expressed in a federal statute and, therefore in conflict with federal supremacy, are held to be preempted by the federal law. When state actions are challenged as preempted under the supremacy clause, the federal courts ask whether such actions stand "as an obstacle to the accomplishment and execution of the full purposes and objectives of Congress."[3] When state actions directly contradict federal law, the federal law prevails and invalidates them.[4] Similarly, when Congress declares that federal power exclusively "occupies the field" of regulation in a certain area, state action that trespasses on the field is invalid even if it complies with the substance or purpose of the federal law.[5] Thus two primary guidelines govern preemption cases: first, in the case of direct conflict between state actions and federal law, federal law prevails; second, where the federal government has "occupied" a particular regulatory field, state action is invalidated.[6]

---

1. E.g., *Phillips Petroleum Co. v. Mississippi,* 484 U.S. 469 (1988).
2. Id.; *Shively v. Bowlby,* 152 U.S. 1 (1894).
3. *Hines v. Davidowitz,* 312 U.S. 52, 67 (1941).
4. *McDermott v. Wisconsin,* 228 U.S. 115, 132 (1913).
5. *Silkwood v. Kerr-McGee Corp.,* 464 U.S. 238, 248 (1984).
6. See, e.g., Tribe, *American Constitutional Law,* 479 (2d ed. 1988).

But these guidelines apply only in clear-cut cases, and preemption law is more complicated than they would indicate. For example, Congress may, and often does, allow considerable scope for state action even though it may have legislated in a particular field. Federal laws, including the Clean Water Act and the Clean Air Act, authorize and, in fact, depend upon state agency action to carry out major components of these federal programs.[7] Other federal laws such as the Comprehensive Environmental Response, Compensation, and Liability Act (CERCLA) authorize the states to exercise considerable authority in regulating in the areas addressed by such federal authority.[8]

Further, the Supreme Court has evolved different tests for particular fields by which to measure state actions. As one legal scholar has succinctly stated:

> The Supreme Court's approach to Supremacy Clause problems differs by subject matter and pays particular respect to the traditions and precedents, both legislative and judicial, of the field under review. The Court is less inclined to find preemptive intent or scope to an act of Congress that involves a field or regulation not traditionally federal and that confronts state laws in areas traditionally left to the states. . . . Other important policy considerations in Supremacy Clause cases are whether there is an apparent need for national uniformity in a federal scheme; whether the federal scheme is a passive command or involves active regulatory efforts of an agency; whether the state laws in question are even-handed or discriminate against the federally-protected interests; and whether other constitutional difficulties presented by a state law can be avoided by according preemptive scope to federal laws.[9]

The Supreme Court has itself noted that "there can be no one crystal clear distinctly marked formula" for applying preemption analysis, but that it necessarily involves a case-by-case determination.[10] Indeed, the Supreme Court has stated that preemption of state regulatory power by federal regulation in a particular area will not be presumed and requires a high degree of proof. Preemption will not operate "in the absence of persuasive reasons—either the nature of the subject matter permits no other conclusion or the Congress has unmistakenly so ordained."[11]

An example of how the Supreme Court applies these considerations was demonstrated in a recent case, *California Coastal Comm'n v. Granite Rock Co.,*[12] in which the Court upheld California's right to review a private mining project occurring on U.S. Forest Service land and to require that the project be permitted under state law. At issue in the case were claims by the mining company and the federal government that several federal laws preempted any state environmental

7. 33 U.S.C. sec. 1251 et seq.; 42 U.S.C. secs. 7401–642.
8. 42 U.S.C. sec. 9601.
9. Cohen, *Handbook on Federal Indian Law,* 272 (1982).
10. *Hines v. Davidowitz,* 312 U.S. 52, 67 (1941).
11. *Florida Lime & Avocado Growers v. Paul,* 373 U.S. 132, 142 (1963).
12. 480 U.S. 572 (1987).

regulation of mining activity taking place entirely upon federal land. The Court closely examined the federal laws that arguably preempted any action by California (including the Federal Land Policy and Management Act (FLPMA),[13] the National Forest Management Act,[14] and the CZMA).[15] With respect to the Forest Service laws and regulations, the Court found that they "not only are devoid of any expression of intent to preempt state law, but rather appear to assume that those submitting [mining plans] will comply with state law."[16] In evaluating FLPMA, the Court found that compliance with state law was also part of FLPMA's regulatory process.[17]

The Court then focused on the alleged conflict between the requirements of the CZMA and the state's permit in the context of the CZMA's explicit exclusion of certain federal coastal lands ("lands the use of which is by law subject solely to the discretion of or which is held in trust by the Federal Government") from the state's coastal zone.[18] In spite of this exclusion, the Court found that Congress did not intend the CZMA to preempt state requirements vis-à-vis federal activities on federal lands: "[b]ecause Congress specifically disclaimed any intention to preempt preexisting state authority in the CZMA, we conclude that even if all federal lands are excluded from the CZMA definition of 'coastal zone', the CZMA does not automatically preempt all state regulations of activities on federal lands."[19]

Based upon the Supreme Court's general preemption analysis, several factors should operate to protect most state actions under the public trust doctrine from federal preemption claims. First, where a state's "historic police powers" are at issue, the Court "start[s] with the assumption that [they] were not to be superseded by the Federal Act unless that was the clear and manifest purpose of Congress."[20] That assumption would seem equally valid with respect to the state's historical public trust responsibilities, because public trust lands and the public rights which attach to them are precisely the kinds of interests that have been traditionally left to the states. Second, the Court will scrutinize the federal statute for indications that Congress did not intend federal regulations to be exclusive[21]

13. 43 U.S.C. sec. 1701. et seq.

14. 16 U.S.C. sec. 1600 et seq.

15. 16 U.S.C. sec. 1451 et seq.

16. 480 U.S. at 573.

17. Id. at 584.

18. 16 U.S.C. sec. 1453(1).

19. Id. at 593.

20. *Hillsborough County,* 471 U.S. 707, 715 (1985); *Ray v. ARCO,* 435 U.S. 151, 157 (1978); *Rice v. Santa Fe Elevator Corp.,* 331 U.S. 218, 230 (1947).

21. See *Granite Rock* (Coastal Zone Management Act does not diminish state authority); see also *Hillsborough County* (Food and Drug Administration stated at time of promulgating regulations that it did not intend them to be exclusive); *Friends of the Earth v. U.S. Navy,* 841 F.2d 927, 934 (9th Cir.

or, to the contrary, required federal agencies to consult with state authorities and to comply with their regulations.[22] Third, even where federal legislation or regulation seemed pervasive and comprehensive, the Court has found that room was left for state regulatory authority in the absence of an actual conflict.[23]

Judicial recognition of the cotrusteeship relationship between the federal and state governments[24] further supports an assumption that the potential for actual conflict between federal and state control over public trust lands and waters is relatively minimal. It can therefore be argued that, unless the language or legislative history indicates an unambiguously clear federal intent to the contrary, a state's historical public trust authority is presumptively not preempted by a federal regulatory scheme or by a particular federal permitting decision.

2. Commerce, Navigation, and Navigability

The public trust activities of states and state coastal agencies very frequently extend to an area of federal power that has involved most of the cases of actual preemption under the supremacy clause, namely the commerce clause.[25] The commerce clause of the Constitution (art. I, cl. 3, § 8) grants Congress the

---

1988) (Clean Water Act specifically acknowledges right of states to regulate discharge of dredge and fill in navigable waters within their jurisdiction); *Chemclene Corp. v. Commonwealth*, 91 Pa. Comm. 316, 497 A.2d 268 (1985) [section 114(a) of Comprehensive Environmental Response, Compensation, and Liability Act contains broad disclaimer of preemptive intent].

22. See *Granite Rock* (National Forest Management Act and Federal Land Policy and Management Act require coordination with state and local authorities and consistency with their plans); see also *Friends of the Earth* (National Defense Authorization Act, Clean Water Act, and Coastal Zone Management Act all require consistency with state and local programs).

23. Where the federal scheme does not indicate an intent to preempt "reasonable state environmental regulation . . . the use of a permit requirement to impose the state regulation does not create a conflict with federal law where none previously existed." *Granite Rock*, 107 S.Ct. at 1429. "To infer preemption whenever an agency deals with a problem comprehensively is virtually tantamount to saying that whenever a federal agency decides to step into a field, its regulations will be exclusive. Such a rule, of course, would be inconsistent with the federal-state balance embodied in our Supremacy Clause jurisprudence." *Hillsborough County*, 471 U.S. at 717. "Thus, if an agency does not speak to the question of preemption, we will pause before saying that the mere volume and complexity of its regulations indicate that the agency did in fact intend to preempt." *Id.* at 718, 105 S.Ct. at 2377; *Granite Rock*, 107 S.Ct. at 1426. See also *Chemclene, supra* (state license requirement for hazardous waste transporters to post bond not preempted by federal statute which provides that transporters who establish evidence of financial responsibility cannot be required by state law to produce additional evidence).

24. See discussion of the *Illinois Central, 1.58 Acres,* and *Alameda II* cases, ch. 4, sec. C.

25. The earliest and most celebrated supremacy/preemption decision of the United States Supreme Court in fact involved a clash of state and federal regulatory power over navigation, which was held to be a species of commerce in *Gibbons v. Ogden,* 9 Wheat. (22 U.S.) 1 (1824), where the Court held that New York statutes granting an exclusive monopoly to use steam navigation on the waters of the state were null and void insofar as they applied to vessels licensed by the federal government to engage in coastal trade.

authority "[t]o regulate Commerce with foreign nations, and among the several states." Described as "the chief source of congressional regulatory power,"[26] the commerce clause permits Congress to exercise extensive authority over the nation's waters, including navigation.[27]

State action, including action under the public trust doctrine, in direct conflict with any federal law enacted pursuant to the federal government's constitutional authority to regulate commerce and navigation would fail, according to the principles pertaining to preemption.

Whether a body of water is navigable, and therefore subject to such plenary regulation and control by the federal government under the commerce clause, is determined according to federal law. Historically, navigable waters were those that were "navigable in fact," a category of waters defined to include waters that "are used, or are susceptible of being used, in their ordinary conditions as highways for commerce, over which trade and travel are or may be conducted in the customary modes of trade or travel on water."[28] This test of "navigability in fact" has been expanded by the Supreme Court to include waters that were navigable at any time in the past[29] and waters that are not navigable but are capable of becoming navigable through "reasonable improvements."[30] The federal government currently defines navigable waters of the United States to include "those waters that are subject to the ebb and flow of the tide and/or are presently used, or have been used in the past, or may be susceptible for use to transport interstate or foreign commerce."[31]

Navigability also determines whether states have title to waters and submerged lands within state boundaries. Unlike the expansive development of the "navigability in fact" criterion in order to determine whether waters are "navigable" for purposes of the commerce clause, navigability for purposes of determin-

26. Tribe, *American Constitutional Law,* 306 (2d. ed. 1988).

27. *Gibbons v. Ogden,* 9 Wheat. 1 (1824). The law with respect to Congress's power of fostering and protection of navigation were early summed up in a frequently cited passage from the Court's opinion in *Gilman v. Philadelphia,* 3 Wall (70 U.S.) 713, 724–25 (1866):

> Commerce includes navigation. The power to regulate commerce comprehends the control for that purpose, and to the extent necessary, of all the navigable waters of the United States which are accessible from a state other than those in which they lie. For this purpose they are the public property of the nation, and subject to all requisite legislation by Congress. This necessarily includes the power to keep them open and free from any obstruction to their navigation, interposed by the states or otherwise; to remove such obstructions when they exist; and to provide, by such sanctions as they may deem proper, against the occurrence of the evil and for the punishment of offenders. For these purposes, Congress possesses all the powers which existed in the states before the adoption of the national Constitution, and which have always existed in the Parliament in England.

28. *The Daniel Ball,* 77 U.S. 557, 563 (1871).

29. *Arizona v. California,* 283 U.S. 423 (1931).

30. *United States v. Appalachian Power Co.,* 311 U.S. 377 (1940).

31. 33 C.F.R. sec. 329.4 (1988).

ing state ownership depends strictly upon whether waters were navigable in their natural and ordinary condition at the time of statehood. The applicability of the public trust to such lands, however, is not necessarily dependent upon the navigability of waters. With respect to tidelands, the states "upon entry into the Union received ownership of all lands under waters subject to the ebb and flow of the tide."[32] (With respect to lands under freshwater lakes and rivers, the applicability of the public trust doctrine does not depend upon ownership but rather attaches to such lands as far as such lakes and rivers are navigable.)[33]

Despite the unquestioned paramount power of Congress over state law in the area of navigation, state regulation of navigation still is afforded substantial leeway when there is no applicable act of Congress involved, no need for a uniform national rule, and no evidence that the state action impedes the free and efficient flow of interstate or foreign commerce.[34] Before such state regulation can be found preempted by federal act, there must be a "repugnance or conflict [that] is so 'direct and positive' that the two cannot 'be reconciled or consistently stand together,'" an obviously stringent standard relatively lenient of state power in the area.[35]

In similar fashion, the Supreme Court has often upheld state legislation permitting the construction across navigable streams of dams,[36] booms, and other shore protections,[37] as well as ruled in support of state legislation authorizing the erection of bridges and the operation of ferries across such streams.[38] Bridges, it is true, may obstruct some commerce, but they may more than compensate for this by aiding other commerce.[39] As the Court noted in *Huse v. Glover*,[40] it should not be forgotten that "the State is interested in the domestic as well as in the interstate and foreign commerce conducted on the [local rivers] . . . and to increase its facilities, and thus augment its growth, it has full power. It is only when, in the judgment of Congress, its action is deemed to encroach upon the navigation of the river as a means of interstate and foreign commerce, that that

32. *Phillips Petroleum Co. v. Mississippi*, 108 S.Ct. 791, 795 (1988).

33. *Barney v. City of Keokuk*, 94 U.S. 324, 338 (1877); *Illinois Central* at 435–36 (1892).

34. Thus, in *Cooley v. Port Wardens*, 12 How. (53 U.S.) 299 (1851), the Court held that states were generally entitled to enact laws adapted to local commercial needs in the absence of conflicting congressional legislation, and that a pilotage law was such a law and was constitutionally applicable to vessels engaged in the coastal trade until Congress acted to the contrary. See also *Ray v. ARCO*, 435 U.S. 151 (1978) (state law requiring tugboat escort of certain vessels in Puget Sound.)

35. *Kelly v. Washington*, 302 U.S. 1, 10 (1937).

36. *Wilson v. Blackbird Creek Marsh Co.*, 2 Pet. (27 U.S.) 245 (1829).

37. *Pound v. Turck*, 95 U.S. 459 (1878); *Lindsay and Phelps Co. v. Mullen*, 176 U.S. 126 (1900).

38. *Gilman v. Philadelphia*, 3 Wall (70 U.S.) 713 (1866).

39. Id. at 729. See also *Escanaba Co. v. Chicago*, 107 U.S. 678 (1883); *Cardwell v. American Bridge Company*, 113 U.S. 205 (1885).

40. 119 U.S. 543 (1886).

body may interfere and control or supersede it. . . . How the highways of a state, whether on land or by water, shall be best improved for the public good is a matter for state determination, subject always to the right of Congress to interpose in the cases mentioned."[41]

The same principle has been applied to the construction of piers and wharves in a navigable stream,[42] as well as to harbor improvements by a state for the aid and protection of navigation;[43] and to reasonable tolls charged for the use of such aids and reasonable regulations laid down governing their employment.[44] Consequently, the preemption risks posed by the commerce clause, while not insubstantial and deserving always to be fully appreciated, appear manageable by properly advised states and their coastal agencies.

3. Sovereign Immunity

While a state may thus assert public trust authority over activities that require a federal permit, as long as there is no direct "repugnance or conflict," the same is not necessarily true with respect to projects undertaken by the federal government itself on public trust lands within a state's borders. Here, the supremacy analysis leads to the opposite result, a presumption of sovereign immunity.[45] This presumption is only overcome by an explicit waiver of such immunity by Congress. Federal projects "are subject to state regulation only when and to the extent that Congressional authorization is clear and unambiguous."[46]

A waiver of sovereign immunity can be broad or narrow. It may require only substantive compliance with state regulation (i.e., meeting its standards) or procedural compliance as well (e.g., applying for permits, record keeping). It may single out specific facilities and activities or generally waive immunity from all "requirements . . . respecting the control of . . . pollution and waste disposal."[47]

41. Id. at 548–49.

42. *Packet Co. v. Keokuk,* 95 U.S. 80 (1877); *Ouachita Packet Co. v. Aiken,* 121 U.S. 444 (1887).

43. *Prosser v. Northern Pacific Railroad,* 152 U.S. 59 (1894); see also *Sands v. Manistee River Imp. Co.,* 123 U.S. 288 (1887); *Gring v. Ives,* 222 U.S. 365 (1912).

44. *Packet Co. v. Keokuk,* 95 U.S. 80 (1877); *Transportation Co. v. Parkersburg,* 107 U.S. 691 (1883).

45. The sovereign immunity doctrine historically bars legal action against a sovereign government unless it consents to such action. Note, however, that only acts of sovereignty are immune. Where the federal government or its agency engages in nongovernmental activity, particularly commercial activity that ordinarily would be considered private in nature, its actions may not be shielded from legal challenge or state regulation. E.g., *McCulloch v. Maryland,* 17 U.S. (4 Wheat.) 316, 426 (1819); *Thompson v. Pacific Railroad,* 76 U.S. (9 Wall) 579, 588 (1869).

46. *Environmental Protection Agency v. California ex rel. State Water Resources Control Bd.,* 426 U.S. 200, 211 (1976); *Hancock v. Train,* 426 U.S. 167, 179 (1976); *Minnesota v. Hoffman,* 543 F.2d 1198, 1209 (8th Cir. 1976), *cert. den'd,* 430 U.S. 977 (1977). See generally, Breen, "Federal Supremacy and Sovereign Immunity Waivers in Federal Environmental Law," 15 Env'l L. Rep. 10326 (1985).

47. See Clean Air Act, 42 U.S.C. secs. 7418 et seq.; Clean Water Act, 33 U.S.C. secs. 1323 et seq.; Resource Conservation and Recovery Act, 42 U.S.C. secs. 6961 et seq.

Where a waiver generally calls for compliance with "requirements," however, the courts have construed this narrowly to mean that federal projects must meet objective and specific standards for pollution control.[48]

Nevertheless, both Congress and the judiciary repeatedly have made clear that some degree of state control over federal activities within its borders is not only permitted but often desirable. Notwithstanding the presumption of sovereign immunity, state regulation is not per se barred by the property, supremacy, or plenary powers clauses in the Constitution.[49] Furthermore, long-standing policies, as manifested in specific acts of Congress, as well as several court decisions, recognize that the federal and state governments share responsibility, in particular with respect to environmental protection and natural resources management. This concept of "dual regulation" provides indirect support for state regulation of federal activity, especially in coastal areas where the state has historic vital interests and at least concurrent authority over public trust land and waters.

It is important, however, to distinguish "federally permitted" from "federally conducted" activity. As long as a state is not clearly preempted, it may regulate federally permitted activities on public trust lands and waters. Exercise of state public trust authority over federal projects, on the other hand, requires a clear and unambiguous waiver of sovereign immunity and is limited by the terms of the waiver. Unless it is expressly provided otherwise, the state is limited to enforcement of clear and objective standards. As is further explained in the next section, the state may be able to regulate, but cannot further restrict, federal projects in the absence of specific congressional authorization.

4. Regulate But Not Prohibit?

In its 1987 *Granite Rock* decision the Supreme Court upheld the California Coastal Commission's claim that a private mining operation on federally held public lands was subject to its consistency review under the Coastal Zone Management Act and required a state permit under the California Coastal Act (CCA), notwithstanding federal approval.[50] The Court reaffirmed that preemption requires a "clear and manifest" federal intent or an actual conflict between federal and state law, and found neither. It has been noted, however, that the holding, while supporting the concept of concurrent authority, was quite narrow.[51] First,

48. See, e.g., *EPA v. California*, 426 U.S. at 218; *Hancock v. Train*, 426 U.S. at 180; see also Breen, 15 Env'l L. Rep. at 10328 and nn. 36–43.

49. See, e.g., *Granite Rock*, 107 S.Ct. at 1425; *Kleppe v. New Mexico*, 426 U.S. 529, 543 (1976); *Hancock v. Train*, 426 U.S. at 179; see also Fairfax and Cowart, "Judicial Nationalism vs. Dual Regulation on Public Lands: Granite Rock's Uneasy Compromises." 17 Env'l L. Rep. 10276 (1987).

50. *California Coastal Comm'n v. Granite Rock Co.*, 107 S.Ct. 1419. For an extensive discussion of the district court, Ninth Circuit, and Supreme Court opinions, see Fairfax and Cowart, *supra* note 49.

51. Fairfax and Cowart, 17 Env'l L. Rep. at 10282–83.

the Court specifically left open the possibility that a future application of the same state permit requirement could in fact conflict with federal law.[52] Second, the Court construed the permit requirement as imposing a "reasonable state environmental regulation" which did "not mandate particular uses for the land but require[d] only that, however the land is used, damage to the environment is kept within prescribed limits."[53] Finally, the Court may have been unwilling to let Granite Rock Company "create" a conflict in authority to avoid regulation.[54] The implication was that the outcome could conceivably be different were the federal government to contest the state's regulatory authority.

A year after *Granite Rock,* however, the Ninth Circuit Court of Appeals rejected a claim by the federal government that sovereign immunity shielded a U.S. Navy construction project from state regulation under Washington's Shoreline Management Act (SMA).[55] The court found that the governing federal statutes did not preempt state regulation but in fact required consistency.[56] Rejecting the Navy's claim that the SMA is essentially a land use law implementing the state's coastal zone management plan, the court held that it regulates and controls dredging and water quality instead, within the scope of the federal Clean Water Act, which waives sovereign immunity.[57] The state's permit requirement therefore applied regardless of whether construction took place on federal or state land, and regardless of the fact that a federal permit was required and had in fact been obtained.[58]

The distinction between land use control and environmental regulation appears to be a mixed blessing for a state taking the comprehensive approach to coastal management. The analysis allows no state interference with federal use of public trust lands or waters unless it can be shown that the state merely seeks to impose a reasonable environmental regulation within the meaning of a federal statute or within the scope of an express waiver of sovereign immunity provided by that statute. With respect to federally permitted activities on public trust lands or waters, a state could, under this view, only impose a reasonable environmental regulation if the governing federal statute requires consistency, or at least does not preempt the regulation, and if the state requirements do not actually conflict

52. 107 S.Ct. at 1432.

53. Id. at 1428–29. The dissent argued that the California law is plainly a land use statute and the permit requirement was clearly preempted as a land use control device. Id. at 1441.

54. See Fairfax and Cowart, 17 Env'l L. Rep. at 10287.

55. *Friends of the Earth v. U.S. Navy,* 841 F.2d 927 (9th Cir. 1988).

56. Id. at 931, 936.

57. Id. at 935.

58. Id. at 936. The U.S. Army Corps of Engineers had issued a permit under section 404 of the Clean Water Act; under section 401 a state must certify that the permitted activity will not adversely affect water quality.

with federal policy or with the methods through which the federal government seeks to attain the policy's objectives.[59]

Instead of relying on this questionable distinction, which may not live long,[60] and rather than trying to make public trust limitations look like environmental regulations, the states should take the position that their historical public trust authority should be recognized independently. A state's knowledge of the nature and value of public trust resources within its borders should at least inform federal permitting decisions and, if necessary, restrict them, in order to guarantee adequate management of those resources consistent with the common law doctrine as well as congressional mandates.

5. Expanding State Authority under the Public Trust Doctrine

Previous parts of this book have discussed the application of the public trust doctrine as an affirmative tool for environmental protection and natural resources management. In its modern form, the doctrine should also create more room for states to assert authority over those uses of public trust land and waters that are subject to federal permits. The states can find support in the Supreme Court's emphasis on the federal-state balance in its supremacy analysis; in the Court's presumption that state police powers are not presumed to be preempted by acts of Congress; and in the explicit recognition of concurrent federal and state authority by both Congress and the Court in the area of environmental regulation. Like environmental protection, management of public trust assets needs national political commitment and policy formulation, as well as substantial control and contribution of expertise by the states. Like its police powers, a state's public trust authority was not "surrendered" upon entry into the federal system. Moreover, management of coastal public trust resources invariably affects public health and safety, traditionally protected under a state's police powers.

The responsibility for coastal zone management is shared by the federal and state governments under the federal Coastal Zone Management Act. This statute clearly mandates full participation by the states in protecting, enhancing, and preserving the assets of the public trust and thus effectively embodies the historical tenets of the public trust doctrine. The CZMA encourages the states, as cotrustees with the federal government, to exercise their "full authority" to protect, preserve, and manage coastal lands, waters, and resources for the benefit of their citizens. In this scheme, the state role is in fact dominant and is not viewed as intrusive. The emphasis is upon shared responsibility and away from conflict.

---

59. A state cannot regulate away a certain use that the federal scheme seeks to encourage.
60. See both the concurring and dissenting opinions in *Granite Rock,* 107 S.Ct. at 1432, 1440.

The appearance of conflict is minimized when a state asserts its public trust authority on the basis of a comprehensive management plan mandated by a state legislature and designed to implement legislatively articulated public trust priorities. Both policy declaration and program implementation must reflect a balancing of competing interests,[61] which will minimize the likelihood of "actual conflict" with federal policy or regulations. Based on an informed balancing of competing interests, a public trust management program should provide "clear and objective standards" for the appropriate uses of public trust land and waters, which should strengthen the state's position under a waiver of sovereign immunity.[62]

In summary, in both its common law and statutory form the public trust doctrine permits the reasonable exercise of state authority even with respect to federally conducted or permitted activities. The doctrine should support a presumption that Congress does not intend to preempt state public trust authority; narrow the likelihood of conflict between federal and state schemes; and broaden the scope of a state's consistency review as well as its authority under a waiver of immunity. When asserting its authority, a state should emphasize the federal-state cotrusteeship and the concurrent or at least consistent exercise of authority it requires. By developing a comprehensive management program, a state can further minimize the likelihood of conflict and at the same time expand the scope of its ability to inform federal agency actions and permitting decisions with an appreciation of the interests protected by the public trust doctrine.

B. ASSERTING THE PUBLIC TRUST VIA FEDERAL CONSISTENCY

This section discusses an important technique available to states in implementing public trust principles vis-à-vis federal agencies and federally permitted projects without violating federal supremacy: the "consistency" provisions of the federal Coastal Zone Management Act, in which Congress disclaimed any intention to preempt state authority in the coastal zone and

61. See, e.g., Mass. Gen. L. chs. 91, sec. 131; Mass. Regs. Code tit. 310, sec. 10.21; N.C.G.S. sec. 113A; N.C. Admin. Code tit. 15r. sec. 7H.0203; Wash. Rev. Code sec. 90.58; Skagit County Shoreline Management Master Program; see also ch. 3, sec. A.

62. Such standards may include: (1) general rules, for example giving priority to conservation, control of pollution, balanced development, or water-dependent uses; (2) "public purpose" rules or exceptions, for example to allow construction and maintenance of public utilities; and (3) more specific rules which focus on particular types of activity or use of particular resources within the management area. The more specific regulations should be based on an inventory of "significant features" each of which may require its own set of rules and priorities, as well as an inventory of traditional or "valid" uses. See Mass. Regs. Code tit. 310, secs. 10.25–10.34 ("resource areas"); Skagit County Shoreline Management Master Program ("shoreline areas"), sec. 6–1 (1983).

required federal agency activities and federal permittees to act consistently with approved state coastal policies. By incorporating public trust principles into their coastal management programs, states can assure avoidance of federal preemption, but also can affirmatively ensure that federal and federally permitted projects do not undesirably affect public trust lands and resources.

A state's public trust interests may be subordinated to federal law, as discussed in the preceding section. Yet public trust principles can be used to limit, modify, or prohibit both the activities of federal agencies and private actions requiring federal permits under the federal consistency doctrine of the CZMA.[63] There are two major components of the consistency doctrine relevant to public trust lands and resources. According to the first, federal agencies must themselves act consistently with the federally approved state coastal programs.[64] Second, private applicants for federal permits for activities affecting coastal lands, waters, and resources must also act consistently with state programs.[65] These requirements are discussed more fully below.

Section 307(c) (1) of the CZMA, as amended in 1990, describes the standard of compliance federal agency activities must meet with respect to state coastal programs.[66] Federal agency activities conducted either in or outside the coastal zone "affecting" the land or water uses or natural resources of the coastal zone must be consistent "to the maximum extent practicable" with the "enforceable" policies of state programs.[67] Federal agencies are required initially to determine the consistency of their own activities with state coastal policies and authorities,[68] and to submit to the state program sufficient information and data to enable the

63. 16 U.S.C. sec. 1456.

64. 16 U.S.C. sec. 1456(c) (1).

65. 16 U.S.C. sec. 1456(c) (3).

66. Although some doubt had been created concerning the effect of section 307(c) (1) on federal activities as a result of the 1984 decision in *Secretary of the Interior v. California*, 464 U.S. 312 (1984), the 1990 amendments to the consistency provisions of the CZMA overturned this controversial decision and restored state consistency review authority over any federal agency activity affecting coastal lands, waters, and resources. The 1984 decision had held that oil and gas lease sales conducted under the Outer Continental Shelf Lands Act (43 U.S.C. secs. 1331–56) were not covered by the federal consistency provisions of the CZMA because, in the Court's opinion, Congress did not intend to include lease sales among the federal agency activities subject to the CZMA. The legislative history to the 1990 amendments expressly identify oil and gas lease sales as subject to the consistency provisions of the CZMA and state that the 1990 amendments allow no exceptions for any federal agency activities to these provisions. *See* Archer, "Evolution of the Major 1990 CZMA Amendments: Restoring Federal Consistency and Protecting Coastal Water Quality," 1 Territorial Sea Journal 191–222 (1991); and Archer and Bondareff, "Implementation of the Federal Consistency Doctrine—Lawful and Constitutional," 12 Harvard Env'l L. Rev. 115–56 (1988).

67. 16 U.S.C. sec. 1456(c) (1). The phrase "to the maximum extent practicable" has been interpreted by the National Oceanic and Atmospheric Administration since 1977 to require "full consistency" unless the federal agency is prevented from meeting this requirement by federal law (15 C.F.R. sec. 930.32).

state to make its own decision.[69] Although mediation by the secretary of commerce is available to settle disputes between a state program and a federal agency concerning the consistency of federal activities, the state's legal recourse is to seek an injunction in federal court to prevent the agency from carrying out an inconsistent activity.[70]

As a result of the 1990 amendments to the consistency provisions, the president may grant an exemption allowing an inconsistent federal agency activity, but only after a federal court has determined the activity to be inconsistent with state coastal policy and the president has declared that, despite the inconsistency nevertheless the activity is necessary in the "paramount" interest of the nation. Even so, only elements of the federal agency activity found by the court to be inconsistent are exempted from consistency—all other elements of the activity must comply with state law. In effect, section 307(c) (1), as amended, constitutes a waiver of immunity with respect to federal agencies whose activities affect coastal lands, waters, and resources.[71]

Section 307(c) (3) of the CZMA sets forth the consistency requirements applicable to private activities requiring federal permits, including outer continental shelf (OCS) oil and gas exploration, development, and production activities.[72] If such an activity in or out of the coastal zone will affect land or water uses (terms that are defined broadly in the CZMA to encompass all activities on coastal lands and waters) or natural resources of the coastal zone, the applicant for a federal permit must "certify" to the federal agency that the activity or project is "consistent" with state coastal programs.[73] If the state objects to the proposed project, the federal agency may not issue the permit unless upon appeal the secretary of commerce overrides the state's objections. Either one of two statutory grounds will support an override by the secretary: the activity, although inconsistent with the state program, is nevertheless determined to be consistent with the national objectives of the CZMA; or the activity is necessary in the interest of national security.[74]

68. 15 C.F.R. sec. 930.33.

69. 15 C.F.R. sec. 930.39.

70. 15 C.F.R. sec. 930.116 (1986). Mediation by the secretary of commerce has not been an effective means of resolving section 307(c) (1) disputes.

71. See discussion of the waiver of sovereign immunity expressed in other federal laws in ch. 4, sec. A, subsection 3.

72. 16 U.S.C. sec. 1456(c) (3) (A), (B) (supp 1993).

73. Id. "Land uses" and "water uses" are defined in 16 U.S.C. sec. 1453(10) and (18).

74. 16 U.S.C. sec. 1456(c) (3) (A), (B) (supp. 1993); 15 C.F.R. secs. 930.120–930.134. The secretary has not yet overridden a state's consistency objection on the ground of national security. Overrides have been granted on the first ground—that the proposed activity meets the national objectives of the CZMA. To make this determination, the secretary must make the findings prescribed by the federal consistency regulations at 15 C.F.R. sec. 930.121:

(a) The activity furthers one or more of the competing national objectives of the CZMA;

In view of the size and number of federal and federally permitted activities and projects either in or affecting the coastal zone (including public trust lands and resources), sections 307(c) (1) and (3) provide important and substantial authority enabling the states to require federal and federally permitted activities to adhere to state policies governing the coastal zone.[75]

If public trust principles are incorporated into state coastal management programs, then both federal agency activities and federally permitted projects affecting coastal or trust lands and resources must be consistent with those principles.[76] Several applications of public trust principles to federal agency activities and federally permitted projects via the federal consistency doctrine suggest themselves. For example, federal agencies planning construction projects in the coastal zone affecting public trust as well as coastal lands and resources may be required to modify or forgo such projects if they adversely affect public interests in trust property. Further, federal agencies seeking to sell coastal (and trust) lands to private parties may find that such lands have been designated for uses by the state trustee consistent with the public trust, thereby limiting the ability of the federal government to dispose of such lands.[77]

(b) The individual and cumulative effects of the activity will not cause adverse effects on the natural resources of the coastal zone outweighing the activity's contribution to the national interest (a costs/benefits analysis);

(c) The activity will not violate any requirements of either the Clean Air Act or Clean Water Act; and

(d) There is no reasonable alternative that would permit the activity to be conducted in a manner consistent with the management program.

See Eichenberg and Archer, "The Federal Consistency Doctrine: Coastal Zone Management and 'New Federalism,'" 14 Ecology Law Quarterly 9–68 (1987).

75. In 1983/84, state programs reviewed approximately 400 federal activities, including OCS oil and gas lease sales (conducted by the Department of the Interior before the 1984 decision in *Secretary of the Interior v. California*), federal highway construction projects, airport construction projects, Department of Defense activities and construction projects, and Army Corps of Engineers' projects. Office of Ocean and Coastal Resource Management, Nat'l Oceanic and Atmospheric Administration, Federal Consistency Study, I-i (April 1985). The coastal states concurred with 372, or 93 percent, of these 400 federal agency activities. During the same period, state programs reviewed approximately 5,500 federally permitted activities for consistency under section 307(c) (3) (A) and 435 plans for OCS exploration and development and production under section 307(c) (3) (B). Id. Nearly 5,000 permits were for dredge and fill activities regulated by the Army Corps of Engineers. Although state programs concurred in the great majority of these permits, a substantial number were subject to modification to comply with state coastal policies, evidenced by the 952 state "objections" recorded by the Army Corps of Engineers during this period. Id.

76. For example, the Cape Cod Commission Act passed recently in Massachusetts explicitly incorporates public trust principles into a regional planning process incorporating fifteen local governments. Because Cape Cod falls within the state's coastal zone, the regional policy plan for Cape Cod will be incorporated into the state's coastal zone management program. Cl. 716, 1989 Mass. Acts, 16(c). Assuming the amendment to the state program is federally approved, all future federal and federally permitted projects will be subject to consistency with the public trust principles built into the act.

77. See ch. 4, sec. C, subsection 2 for a discussion of this example.

With respect to federally permitted projects, public trust principles may be used, through the state's consistency review authority, to require project modifications or to prohibit such projects if they unacceptably affect trust lands and resources. Thus Army Corps of Engineers section 404 dredge and fill permits for activities on trust lands or waters, or National Pollutant Discharge Elimination System permits for discharges of pollutants affecting trust property, may be considered in light of use priorities set by the trustee to protect the public's rights in such property.[78]

Coastal managers have gained considerable experience over many years in reviewing a wide range of federally conducted and permitted activities for consistency with state coastal policies. The novelty they may experience in asserting their consistency authority with respect to public trust rights affected by such activities stems from the fact that coastal managers have tended to neglect the public trust doctrine in implementing coastal management programs. Invoking public trust policies in performing consistency reviews, however, should be treated in exactly the same manner as any other coastal policies on which coastal managers may rely to object to a proposed federally conducted or permitted activity. As stated above, the well-advised coastal state should first expressly incorporate public trust principles, as they apply within the state, into its coastal program in order to claim such principles as policies with which the federal agency or permittee must comply. Thereafter, the consistency review would follow the procedures well known to federal and state agencies, as described briefly above. Coastal managers would actively consider public trust policies in reviewing projects, negotiate with federal agencies and permittees to obtain modifications to such projects on the basis of the public trust doctrine, and in certain cases, object to projects on the ground of inconsistency with state public trust policy. Disagreements regarding the application by the state of its consistency authority to require compliance with public trust policies would be resolved by the federal courts in the case of federal agency activities and by the secretary of commerce in the case of federally permitted projects.

Because of the overlap between the "coastal zone" (as defined by the CZMA) and public trust lands, waters, and resources, the federal consistency doctrine, as described above, allows state programs to use public trust principles to good effect in reviewing federally conducted or permitted activities. However, without the authority provided by the federal consistency doctrine over federal agency activities and federally permitted projects, state public trust law may either be preempted or severely limited in its effect upon such activities. Used together, both doctrines provide state programs with an effective mechanism to enforce their coastal management policies.

78. See generally ch. 3, sec. A.

C. THE FEDERAL ROLE IN THE PUBLIC TRUST DOCTRINE

This section reviews several ways the federal government has influenced the public trust doctrine in the states, then specifically addresses the issues which arise when the federal government itself owns shorelands that have historically been or arguably are impressed with public trust obligations. The questions whether there is a federal public trust doctrine and whether federal ownership of former state lands extinguishes the trust are discussed, and the position is propounded that states should presume that the federal government is as bound to honor the public trust in tidelands within its control as is the state, in the absence of clear congressional intent to the contrary.

1. Generally

The federal government has been an important, if somewhat uncertain, factor in the public trust doctrine's development on these shores. The primary reason for this uncertainty has been the belief, which took root early in our history and law, that the traditional public rights in tidelands and submerged lands (navigation, fishing, and fowling) were fundamentally matters of state rather than national concern, because the former sovereignty of the British Crown over these trust lands had passed to the original states and the people after the Revolution. Yet, despite this widely shared view, the several branches of the federal government have each played at times prominent roles in shaping the public trust doctrine in the United States. The following examples illustrate these roles and their importance.

In enacting the Northwest Ordinance in 1787, providing for the governance of the western territories ceded to the national government by several of the original states, the legislative branch codified the principle that new states carved from this territory would enter the Union on an equal footing with the original states. By so doing, the legislative branch established the principle, critical to public trust theory, that new states would assume title to such trust lands for the benefit of their people in the same manner that the original states became owners and trustees of trust lands for their citizens.[79] As a result, public trust principles have become embodied in the law of all states.

Before new states were created from these ceded lands or from other acquired or conquered lands in the West, including Alaska and Hawaii, the federal government and its agencies ruled them. However, the executive and legislative branches of the federal government were constrained (although not absolutely) in their treatment of public trust tidelands and submerged lands in these territories. According to the federal courts, the federal government exercised a "tempo-

79. Northwest Ordinance of 1787, 1 Stat. 50 (1789).

rary trusteeship" over such public trust lands and was obligated to pass such lands intact to the new states upon their gaining statehood.[80]

By far the greatest influence of the federal government upon the development of the public trust doctrine in the United States has been the series of major Supreme Court decisions, beginning relatively early in our history, interpreting and applying public trust principles in the context of actions by state legislatures or agencies to restrict or destroy public rights in trust lands. This federal court supervision of the state trustees has effectively preserved public trust lands, by invalidating or limiting attempted conveyances of such lands to private owners free of trust obligations and by establishing basic principles governing the conduct of state legislatures and agencies in administering their trusteeship responsibilities.[81]

2. The Public Trust Doctrine and Federal Shorelands

In states in which the federal government owns sizeable tracts of coastal property, state coastal managers interested in implementing public trust principles may be uncertain as to the scope and applicability of those principles with respect to such lands, wholly aside from issues of supremacy and premption. Whether there is an effective federal public trust doctrine, or any public trust doctrine at all, applicable to federally held shorelands (that is, tidelands and submerged lands that, if held by state government, would be subject to the public trust doctrine) is a question that has been considered by only a few lower federal courts and has not reached the United States Supreme Court.[82] Among the few courts that have considered the issue (in the context of federal condemnation of state shore property for federal use) there is a clear difference in opinion.

In *United States v. 1.58 Acres of Land,*[83] the court framed (and answered) this question in the following manner:

> The issue is whether or not the United States may take a full fee simple title to land below the low water mark without destroying the perpetual public trust imposed upon that land. If it cannot, then serious constitutional and statutory questions are raised

80. *Pollard's Lessee v. Hagan,* 3 How. (44 U.S.) 212 (1845).

81. See ch. 1, sec. C, subsection 3. The significant component of the federal role played by federal constitutional authority in operating to define and limit the scope of state action pursuant to the public trust doctrine, principally through the supremacy and commerce clauses of the United States Constitution and the doctrine of preemption, has been discussed in ch. 4, sec. A.

82. The Supreme Court has said both that the states owned the shores of navigable waters and the soil under them [*Pollard's Lessee v. Hagan,* 3 How. (44 U.S.) 212, 223 (1845)]; and also that "all the navigable waters of the United States . . . are the public property of the nation, and subject to all requisite legislation by Congress" [*Gilman v. Philadelphia,* 3 Wall (70 U.S.) 713, 724–25 (1866)], arguably thereby endorsing, if only implicitly, the concept of state-federal "cotrusteeship" under the public trust doctrine discussed below in this section.

83. 523 F. Supp. 120, 122 (D. Mass. 1981).

concerning the power of the federal government to destroy forever an important aspect of the Commonwealth [of Massachusetts] sovereignty. We hold, however, that the United States may obtain full fee simple title to land below the low water mark without destroying the public trust which is administered by both the federal and state sovereigns.

In *United States v. 11.037 Acres*,[84] the court reached a different result:

The crucial issue in this case is whether the United States' condemnation of the 11.037 acres of land extinguishes the State of California's public trust easement in the lands. Because this Court finds that the United States' power of eminent domain is supreme to the State's power to maintain tidal lands for the public trust, the Court concludes that the United States' condemnation of these lands extinguishes the state's public trust easement.

Because of the dearth of legal authority in this area, and the significance of the issues involved, these two cases are worthy of extended discussion.

In *1.58 Acres*, the Coast Guard condemned a small parcel of land to use in connection with its support facilities located on the Boston waterfront. Massachusetts objected that the United States could not obtain a fee simple absolute in the portions of the property lying below the low watermark in Boston Harbor, because such a fee would "vitiate the perpetual public trust that is impressed upon land below the low water mark and which is administered by the Commonwealth."[85] Massachusetts argued that the terms of the federal condemnation declaration allowing such uses of the property "as may be authorized by Congress or by Executive Order" would put the submerged land "forever beyond the state's control for purposes of the trust,"[86] potentially leading to private ownership and use of the trust land in the event the United States later conveyed the property to a private owner. The United States asserted that because the use it intended for the property is a "public use," thereby fulfilling the constitutional and statutory requirement necessary in order to exercise the government's power of eminent domain, nothing remained for the court to review in the federal government's condemnation action.

Normally, the existence of "public use" of the condemned property would be sufficient grounds to conclude the court's review of the federal government's action. Here, however, the public trust which attached to the property compelled the court to examine both the nature of the interest acquired by the federal government in the submerged land and any obligation upon the government as a consequence of the land's public trust burdens. The court briefly reviewed the development of the public trust doctrine from Roman and English common law,

84. 685 F. Supp. 214, 216 (N.D. Cal. 1988).
85. *1.58 Acres* at 121.
86. Id.

noting that the division of the sovereignty of the British Crown after the American Revolution, between the national and state governments, compelled a number of modifications of the common law public trust theory to adapt the doctrine to its new American setting. The court found the more important of these modifications in the best known of the Supreme Court's statements of the American public trust doctrine:

> [I]t is the settled law of this country that the ownership of and dominion and sovereignty over lands covered by tide waters, within the limits of the several states, belong to the respective states within which they are found, with the consequent right to use or dispose of any portion thereof when that can be done without substantial impairment of the interest of the public in the waters, and subject always to the paramount right of Congress to control their navigation so far as may be necessary for the regulation of commerce with foreign nations and among the states.[87]

Thus, according to the Massachusetts federal district court, the Supreme Court in *Illinois Central* has recognized "the division of sovereignty between the state and federal governments" with respect to the ownership and management of public trust lands:

> Those aspects of the public interest in the tideland and the land below the low watermark that relate to the commerce and other powers delegated to the federal government are administered by Congress in its capacity as trustee of the *jus publicum*, while those aspects of the public interest in this property that relate to nonpreempted subjects reserved to local regulation by the states are administered by state legislatures in their capacity as co-trustee of the *jus publicum*.[88]

The concept of "cotrusteeship" enunciated here with respect to the "governmental *jus publicum*" (or the complex of public rights attached to the land of which the government is the trustee) was further elaborated by the court with respect to the *jus privatum*, or ownership interest in the land:

> Since the trust impressed upon this property is governmental and administered jointly by the state and federal governments by virtue of their sovereignty, neither sovereign may alienate this land free and clear of the public trust. When the federal government takes such property by eminent domain, however, the federal government obtains the fullest fee that may be had in land of this peculiar nature: the *jus privatum* and the federal government's paramount *jus publicum*.[89]

The result of the court's formulation of the dual-sovereign nature of the public trust respecting shorelands acquired by the federal government is a limitation upon the federal government's ability to terminate its trust obligations when

---

87. *1.58 Acres* at 123, quoting *Illinois Central,* 146 U.S. 387, 435 (1892)].
88. Id.
89. Id. at 124.

it sells trust lands: "[T]he federal government is as restricted as the [states] in [their] ability to abdicate to private individuals its sovereign *jus publicum* in the land."[90] The district court found this "restriction" in the Supreme Court's decision in *Illinois Central:*

> The trust devolving upon the State [or the federal government] for the public, and which can only be discharged by the management and control of the property in which the public has an interest, cannot be relinquished by a transfer of the property. The control of the State for the purposes of the trust can never be lost. . . . The State can no more abdicate its trust over property in which the whole people are interested, like navigable waters and the soils under them, so as to leave them entirely under the use and control of private parties . . . than it can abdicate its police powers in the administration of government and the preservation of the peace.[91]

Despite its conceptualization of a governmental cotrusteeship with respect to public trust lands, which arises from the split in sovereignty between the states and the federal governments after the American Revolution, the Massachusetts district court explicitly recognized that the state's trusteeship and administration of the lands subject to the trust are in some areas subject to the "paramount rights" of the federal government "to administer its trust with respect to matters within the federal power," particularly matters subject to the federal powers over commerce and navigation.[92]

This view of the public trust doctrine as binding upon federal property was not shared by a federal court in California, however. In *11.037 Acres,* the state of California conveyed to the city of Oakland trust lands that were subsequently filled and used in connection with the administration of a port. In 1983 the federal government condemned the filled lands. California entered the condemnation proceedings asserting that the federal government may only take the trust lands subject to the "public trust easement" of the state. The federal government argued: that because the lands were filled lands at the time of condemnation, the public trust did not apply; or, alternatively, that the condemnation action by the federal government pursuant to the government's constitutional right of eminent domain extinguished all other interests in the land unless specifically acknowledged by the federal government.

On the first legal issue—whether filling tidelands extinguishes the public trust—the court easily found, on the basis of well-established California law,[93] that lands subject to the action of the tides at the time of statehood remain tidelands for the purposes of the public trust after such lands have been filled.

90. Id. at 125.
91. Id. at 124, quoting *Illinois Central* at 452–53.
92. Id. at 124.
93. Citing *City of Long Beach v. Mansell,* 3 Cal.3d 462, 91 Cal. Rptr. 23, 476 P.2d 423 (1970).

Only the state legislature may free such lands from the public trust restrictions that they bear.[94] Not finding any evidence that the California legislature had consented to the destruction of the public trust with respect to these lands, the court rejected the federal government's argument. Concerning the second issue—whether the condemnation action by the federal government extinguished the public trust—the court in *11.037 Acres* believed itself compelled by the supremacy clause of the Constitution to uphold the federal government's action: "The crucial issue in this case is whether the United States' condemnation of the 11.037 acres of land extinguishes the State of California's public trust easement in the lands. Because this Court finds that the United States' power of eminent domain is supreme to the State's power to maintain tidal lands for the public trust, the Court concludes that the United States' condemnation of these lands extinguishes the State's public trust easement."[95]

California argued in this case that the public trust is an "attribute of sovereignty" upon which the United States may not infringe even though it may condemn the land and take title to it. But the court in *11.037 Acres* acknowledged no distinction between title to the trust land and the public trust attaching to it for purposes of the condemnation action: "It is undisputed that the United States may take title to these tidelands through eminent domain proceedings. No reason exists for treating the 'public trust attribute' of sovereignty different from the 'title attribute.'"[96]

Finding that the power of eminent domain is essential to the sovereignty of the federal government, the court refused to "impress the California public trust easement on the lands acquired by the United States," because to do so would subjugate the federal government to the interest of the state. The court acknowledged the contrary position taken in *1.58 Acres* but declined without explanation to follow that decision because of its view of the supremacy clause, which it implied had been erroneously applied by the Massachusetts federal court.[97]

The approach reflected in *1.58 Acres* seems better reasoned and more consistent with the current tenor of federal-state relations. In fact, in another case in the Northern District for California a few years earlier, *City of Alameda v. Todd Shipyards Corp.*,[98] involving two separate decisions, the court reached conclusions that are inconsistent with the holding in *11.037 Acres*. In *Alameda I*, the United States had acquired the greater part of trust lands in 1930 by purchase from the city of Alameda, which had received the lands earlier from the state

94. Citing *Marks v. Whitney*, 6 Cal.3d 251, 261, 98 Cal. Rptr. 790, 491 P.2d 374 (1971).
95. *11.037 Acres* at 216.
96. Id. at 217.
97. Id. at 217.
98. 632 F. Supp. 333 (N.D. Cal. 1986) [*Alameda I*] and 635 F. Supp. 1447 (N.D. Cal. 1986) [*Alameda II*]. Surprisingly, neither decision was cited or mentioned by the *11.037 Acres* judge.

subject to the express legislative limitation that the lands could not be conveyed into private ownership, although the lands could be conveyed to the United States. Subsequently the United States condemned a small parcel of these trust lands that the city had mistakenly failed to convey to it in 1930. In 1970 the United States sought to sell the land to the privately owned shipyard. The court found that the terms of the conveyance by the California legislature to the city were also effective against the United States and would invalidate the attempted sale:

> [The limitation] bars both the City and the United States from conveying the land to a private person. The City of Alameda did not acquire from the state the right to sell the land to a private person, and hence the City could not in turn convey this right to the United States.[99]
>
> In the present instance, the legislature granted the lands to Alameda "in trust." Thus the legislature clearly intended that public trust restrictions would remain. The City did not then have the power to remove the trust restrictions merely by conveying the land to the United States, and the United States did not have the power to remove the trust restrictions merely by conveying the land to Todd Shipyards.[100]

The court in *Alameda I* made it clear that the condemned portion of the trust lands held by the federal government was as much subject to the public trust as the lands sold to the United States by the city:

> The sliver of land the United States acquired by condemnation also carried with it public trust obligations [citing *1.58 Acres*]. Article X of the California Constitution establishes a trust in the land. Even if the land was filled at the time of the condemnation action, the United States still could acquire only the interest of its predecessors in title, the City and the State. Since the State and the City held the land subject to the public trust, the United States could take the land only subject to the public trust.[101]

In *Alameda II*, the court addressed motions to reconsider its decision in *Alameda I*. The federal government argued that the decision was in conflict with the Supreme Court's ruling in *Surplus Trading Co. v. Cook* concerning the condemnation by the United States of state trust lands, where the Supreme Court said:

> It is not unusual for the United States to own within a state lands which are set apart and used for public purposes. Such ownership and use without more do not withdraw the lands from the jurisdiction of the state. On the contrary, the lands remain part of her territory and within the operation of her laws save that the latter cannot affect the title of the United States or embarrass it in using the lands or interfere with its rights of disposal.[102]

99. *Alameda I* at 337.
100. Id. at 339–40.
101. Id. at 341.
102. 281 U.S. 647, 650 (1930).

The district court rejected this argument, on the ground that in *Alameda I* the state had consented to the transfer of the land, unlike the situation in *Cook*.[103] In the former circumstances, the United States takes the land subject to any restrictions in the deed or granting statute, and "the land remains subject to the state laws in effect at the time of the transfer of jurisdiction."[104]

In response to the claim by the United States that it had condemned a portion of the trust lands, and therefore that the federal government had acquired this land free of the public trust, the court apparently created a new test to determine whether public trust interests survive the condemnation action. Citing *1.58 Acres,* the *Alameda* court stated that "[b]y condemnation, the United States acquired both the *jus privatum* (the bare ownership interest) and the *jus publicum* (the public trust interest). . . . The United States may not abdicate the role of trustee for the public when it acquires land by condemnation":

> The United States acquired sole trusteeship of the condemned portion of the land. If this portion was subject to the action of the tides at the time of condemnation, then the United States acquired this portion subject to the public trust, and the United States may not convey this portion to a private party (citing *Illinois Central*). If this portion of the land was not subject to the action of the tides at the time of condemnation, [then] the United States may convey this portion to a private party.[105]

The *Alameda II* court implicitly based this test as to the existence of a federal common law public trust doctrine applicable to the United States government upon its view that there is a governmental trusteeship with respect to tidelands effective against both the state and federal governments. The source of this formulation of a federal common law public trust doctrine may be traced back through *1.58 Acres* to the Supreme Court's decision in *Illinois Central,* and this fruitful concept leads to the imposition of cotrusteeship duties upon both governments. Thus, according to this formulation of the common law public trust doctrine, in circumstances where the United States either acquires public trust shorelands from a willing seller or condemns such lands, the trust survives and operates to restrict subsequent transfers of the land to private parties.[106]

103. *Alameda II* at 1449.
104. Id. at 1449–50.
105. Id. at 1450.
106. Federal courts have rarely addressed the existence of a federal common law public trust outside of condemnation cases. In *In re Stewart Transportation Co.,* 495 F. Supp. 38 (E.D. Va. 1980), the court refused to dismiss an action by Virginia and the federal government against a barge owner for damages to migratory fowl as a result of an oil spill. The barge owner claimed that neither the state nor the federal government owned the wildlife, and could not sue in the capacity of property owners. The court agreed that neither government owned the birds, but stated "[u]nder the public trust doctrine, the State of Virginia and the United States have the right and duty to protect and preserve the public's interest in national wildlife resources. Such right does not derive from ownership of the resources but from a duty owing to the people." Id. at 40. The opinion provides little analysis of the doctrine to support its conclusion.

On the basis of the few cases that have considered this matter, it is not certain to what extent a federal public trust doctrine would restrict the uses to which the lands could be put.[107] It does seem, however, that the public trust surviving with respect to federally owned shorelands would not prohibit the Congress from extinguishing the public trust in the same manner in which a state legislature might act to extinguish it.[108] This feature of the federal public trust was implicitly recognized by the *Alameda II* court when it observed that, because "Congress has not acted to remove the restrictions that applied to California tidelands when the United States acquired the [condemned] land," the lands were still imbued with the trust.[109]

Whether the Congress would be subject to the restrictions that the Supreme Court since *Illinois Central* has applied to state legislatures when trust lands are conveyed into private ownership and trust rights are arguably extinguished is not clear; but it seems reasonable that both governmental trustees should be subject to the same standards of explicitness with respect to conveyances that might effectively terminate such important public interests.[110] If there is a federal common law public trust doctrine to be found in *Illinois Central* and its progeny, there would appear to be no barrier to imposing such restrictions upon the national as well as the state legislatures. At the very least, coastal managers and their allies concerned with optimal public trust enforcement should realize that, despite the relative paucity of law on the subject, they are justified under existing authorities in operating on the premise that both state and federal governments are presumptively bound to honor the public trust in any shorelands they control, in the absence of clear evidence of congressional intent to the contrary.

The United States government owns large tracts of coastal lands that, if owned by state governments or individuals, would be subject to public trust rights. These lands were either acquired by purchase or condemnation by the federal government, or retained in federal ownership and never conveyed to newly created state governments. A hypothetical case (which example may well become quite common with the lessening of international tension and the anticipated distribution of a "peace dividend") may serve to illustrate the public trust concepts discussed above: the federal General Services Administration (GSA) announces a proposed sale of "surplus" coastal lands. Pursuant to the state's public trust doctrine, the coastal program has prescribed certain uses of shorelands currently owned by the federal government in the event that ownership is relin-

107. Both *1.58 Acres* and the *Alameda I* and *Alameda II* cases recognize that the public trust attributes of the land are subject to the paramount rights surrendered to the federal government by the states in the Constitution.

108. See ch. 2, sec. F.

109. *Alameda II* at 1450.

110. See ch. 2, sec. F.

quished. What action can the state take if the GSA entertains offers inconsistent with the coastal program's policies or maintains that the public trust either never existed or has been extinguished with respect to such lands? Assuming that the state has incorporated public trust principles as enforceable policies in its coastal management program, what course may coastal managers take to ensure that the "sale" is carried out consistently with state coastal public trust policies?

Arguing the precedent of *1.58 Acres* and *Alameda I and II,* the state should (as early and publicly as possible, so that all potential purchasers are on notice) assert that the public trust attached to the land throughout the period of federal ownership, especially in circumstances in which the federal government acquired the land by purchase or condemnation—circumstances specifically addressed in these cases. The state would argue, in effect, that the land was always impressed with the public trust and that upon relinquishment by the federal government, the latent public trust rights spring back in full force. Further, the state should argue that the federal government is itself a cotrustee of the lands and must conduct itself in accordance with the requirements and duties of the trusteeship, including the duty to consider whether sale of trust lands is proper in the first instance. In short, the federal government is obligated to conduct the sale pursuant to the trust and to take into account the public trust responsibilities of its cotrustee—the state.

If the land had remained in federal ownership and had never been conveyed to the state at the time of statehood, coastal managers might frame their argument differently. Drawing upon earlier precedent, they can assert that the "temporary trusteeship" held by the federal government over territorial lands ultimately passed to the state upon statehood continues to apply with respect to lands retained in federal ownership after statehood.[111] Here again the concept of "trusteeship" to define the duties of the federal government respecting such lands establishes norms that the GSA should consider in any proposed "surplusing" action. Is the sale consistent with the government's trusteeship over and the public rights in the lands? Should the lands be conveyed to the state to establish the state's sole trusteeship (perhaps a more effective arrangement putting such lands on a par with other state-owned coastal trust lands)? Should the property be conveyed first to the state and then reconveyed into private ownership, but now clearly subject to public trust rights? Recognition of the federal government's trust responsibilities in these circumstances would help to clarify the public trust policy issues necessarily involved in the disposal of such lands.

In addition to legal actions, a state can exert considerable leverage in negotiations with the GSA over a proposed sale of surplus federal public trust lands. For

111. *Pollard's Lessee v. Hagan,* 3 How. (44 U.S.) 212 (1845).

example, the federal government may choose to make surplus lands available to states for public purposes. A state can usually bring to bear a certain degree of political pressure upon federal agencies in this regard, particularly in cases where important local or state public interests may be served by the conveyance of federal lands to state or local government. Sales to private owners may also be conditioned upon a recognition by purchasers that the lands are conveyed subject to the public trust—a voluntary recognition by the federal government and the purchaser that the public trust does, in fact, spring back upon its sale to private owners. Although the GSA would typically wish to receive the full market value for surplus lands in such cases, nevertheless the facilitation of important public interests of state or local governments may more than offset any losses to the federal treasury.

Coastal states can make such claims to oppose the sale of surplus federal coastal lands independent of their consistency authority under the CZMA. But coastal managers may also make effective use of their federal consistency authority to review such sales.[112] Although one federal court has ruled that the act of selling surplus lands does not trigger the state's consistency review power under the CZMA,[113] another court has found that the leasing of a naval shipyard could be reviewed by the state under its consistency authority.[114] The public trust doctrine was not implicated in either decision. But if the proposed sale involves lands arguably burdened with the public trust and if the state has incorporated coastal public trust policies in its program, it seems reasonably certain that the argument that the sale itself affects the coastal zone would have greater force.

In any event, coastal managers must argue as strongly as possible the grounds for the effects on the coastal zone stemming from the sale of surplus public trust lands in order to invoke their consistency authority over such sales. Although the federal consistency regulations permit the federal agency initially to determine whether such sales are consistent "to the maximum extent practicable" with state coastal public trust policies,[115] coastal managers may exact "full compliance" with their coastal policies, pursuant to the same regulations.[116] Again, asserting state consistency review over the sale of surplus federal lands arguably imbued with the public trust should proceed in the same manner as the review of any

112. See ch. 4, sec. B for discussion of the interplay between the public trust doctrine and the federal consistency doctrine.

113. *Ono v. Harper,* 592 F. Supp. 698, 700 (D.C. Haw. 1983): "mere transfer of title does not directly affect the coastal zone or the state's management program. It does not change the way in which the land is being utilized."

114. *City and County of San Francisco v. United States,* 443 F. Supp. 1116, 1128 (N.D. Cal.), *aff'd,* 615 F.2d 498 (9th Cir. 1980).

115. 15 C.F.R. sec. 930.32.

116. Id.

other federally conducted activity affecting the coastal zone. In short, the fact of former federal ownership should pose no insuperable barrier to states in their efforts to protect public trust interests to the fullest extent.[117]

117. See ch. 4, sec. B.

V
Evolution of a State Public Trust Doctrine Program: Massachusetts Chapter 91

One of the principal lessons to be drawn from this study is that the public trust doctrine is not self-executing. Even a judicially well-articulated public trust doctrine is of limited value if there is no state official or agency explicitly empowered and willing to implement and enforce it. Thus a state desiring a viable and effective public trust doctrine must both articulate the doctrine's principles and create a mechanism for applying those principles so as to achieve an optimal balance of public and private interests in coastal property. This mechanism must resolve disputes between these competing interests, as well as between conflicting public interests in coastal resources, in an open and systematic process.

In most states, the courts were the first, and for decades the only, mechanism available for resolving these disputes, and did so on a case-by-case basis. Because court are limited by the specific facts of the case before them, this ad hoc system often led to inconsistent results and gaps in the doctrine's application to the use of coastal resources. Therefore, legislative and administrative efforts are necessary if a state wants to manage its coastal resources comprehensively and consistently.

In Massachusetts, the judiciary, the legislature, and various administrative bodies have alternated over the last three centuries in taking the lead to develop and apply the public trust doctrine, but with limited overall coordination. Even though Massachusetts boasts the first American statutory codification of the public trust doctrine, a rich history of judicial pronouncements in the area and a statutory tidelands licensing scheme over a century old, it is only in the last twelve years that this solid but fragmented legal base for the public trust doctrine has been codified and coherently restructured so that the doctrine can be effectively and consistently implemented in Massachusetts with maximum public benefit and minimum private disruption.

For these reasons, the evolution of the public trust doctrine in Massachusetts from a one-paragraph description in the colonial Ordinances of 1641 and 1647 to the sixty-page comprehensive regulatory program adopted by the Department of Environmental Protection in 1990 provides an example that should prove instructive for other states that have yet to realize the promise inherent in the doctrine. The Massachusetts program is not perfect and will differ substantively in many ways from that in other states. Still, this case study, in conjunction with the insights and observations discussed earlier, usefully illustrates the key factors and processes that must be addressed for the successful implementation of the public trust doctrine in other states.

A. INTRODUCTION OF THE DOCTRINE INTO MASSACHUSETTS AND JUDICIAL INTERPRETATION

The public trust doctrine was introduced into Massachusetts during the early stages of English colonization as part of the transplanted English common law.[1] Because the English king could not convey *jus privatum* free from the *jus publicum,* the grants to the colonists were, as in England, subject to the *jus publicum.*[2] But the Massachusetts Bay Colony, dependent on maritime trade for its economic health, soon felt obliged to depart from the common law in an effort to stimulate commerce by encouraging private wharf construction. The result of this effort was the Ordinances of 1641 and 1647.[3] The ordinance stated in part:

> Every Inhabitant who is an housholder shall have free fishing and fowling in any great ponds, bayes, Coves and Rivers, so far as the sea ebbs and flowes, within precincts of the towne where they dwell, unless the freemen of the same Town or the General Court have otherwise appropriated them. . . . [I]n all *Creeks, Coves* and other places, about and upon *Saltwater,* where the Sea ebbs and flowes, the proprietor of the land adjoining, shall have propriety to the low-water mark, where the Sea both not ebb above a hundred Rods, and not more wheresoever it ebbs further. Provided that such proprietor shall not by this liberty, have power to stop or hinder the passage of boats or other vessels, in or through any Sea, Creeks, or Coves, to other mens houses or lands.[4]

1. See discussion of early history of doctrine in the United States in ch. 1, sec. C. See also Rice, *Final Report: A Study of the Law Pertaining to the Tidelands of Massachusetts,* 1970, House No. 4932, at 19 (1971). See also *Home for Aged Women v. Commonwealth,* 202 Mass. 422, 89 N.E. 124 (1909); *Commonwealth v. Roxbury,* 75 Mass. 451 (1857).

2. *Commonwealth v. Roxbury,* 75 Mass. 451 (1857).

3. *The Book of the General Lawes and Liberties* at 50, cited in *Opinion of the Justices,* 365 Mass. 681, 685 (1974).

4. Id. Grants under the ordinance conveyed fee simple title in tidelands to the extreme low watermark or 100 rods from the mean high watermark, whichever measure was furthest landward. *Storer v. Freeman,* 6 Mass. 435 (1810). Massachusetts courts have declared that the same body of law applies throughout the commonwealth with respect to tidelands, including areas which were not within the jurisdiction of the Massachusetts Bay Colony when the Ordinance of 1641–47 was enacted. Rice, note 2 above, at 22 (citations omitted).

The proprietary grant of tidelands nonetheless remained subject to the traditional public rights of fishing, fowling, and navigation.[5]

Until 1860 the courts were the sole interpreter of the colonial ordinances. Generally, the courts recognized that the purpose of the ordinances was to encourage and facilitate the building of wharves. As a result, the courts allowed such construction even when it might diminish the common rights of navigation and fishing.[6] The courts explicitly stated that upland owners who built a wharf or enclosure to the low watermark could legitimately exclude public use of the wharf and underlying tidelands.[7] Without such action by the owner, however, the public's rights to navigation, fishing, and fowling could be freely exercised despite the upland owner's proprietary interest in the tidelands.[8]

B. LICENSING LAW LIMITS UPLAND OWNER'S USE OF PUBLIC TRUST UNDER ORDINANCE

The courts' support of upland owners' exclusion of public use of public trust land under certain circumstances was frequently criticized as being at odds with the common law and the reservation of the public's common rights in the colonial ordinances.[9] Nevertheless, the practice of excluding the public from filled tidelands was affirmed in later judicial decisions.[10] Furthermore, during the early 1800s, the commonwealth passed a series of statutes designed to enable private landowners to construct and maintain wharves, purporting to convey tideland interests seaward of the low watermark to private landowners.[11] Like the colonial ordinances and wharfing statutes in many states,[12] these Massachusetts statutes encouraged private landowners to profit while providing a benefit (the wharves and their facilitation of commerce) to the public. In practice, the combined private profit and public benefit stimulated such rapid development in

5. *Opinion of the Justices,* 365 Mass. 681, 685 (1974). See, e.g., *Butler v. Attorney General,* 195 Mass. 79, 83, 80 N.E. 688–89 (1907); *Commonwealth v. Alger,* 61 Mass. 53, 70–71 (1851); *Arundel v. M'Culloch,* 10 Mass. 70 (1813).

6. Rice, note 2 above, at 23–24. See, e.g., *Commonwealth v. Charlestown,* 18 Mass. 180, 183 (1822).

7. *Austin v. Carter,* 1 Mass. 231–32 (1804).

8. Id.

9. Rice, note 2 above, at 24. See also *Alger,* 61 Mass. at 73–74.

10. See, e.g., *Alger,* 61 Mass. 53 (1851).

11. See, e.g., 1836 Mass. Acts c. 45; 1834 Mass. Acts c. 115; 1833 Mass. Acts c. 35; 1832 Mass. Acts c. 57, 102; 1830 Mass. Acts c. 96, 113; 1829 Mass. Acts c. 92; 1816 Mass. Acts c. 23; 1815 Mass. Acts c. 171.

12. See discussion of riparian rights to exclude the public under such statutes in ch. 3, sec. C, subsection 3.

Boston Harbor that by 1835 the legislature felt that regulation of harbor development was necessary.[13]

In response, the legislature appointed three commissioners to survey Boston
Harbor in order to establish harbor lines to safeguard public rights in the harbor.[14] Relying on the commissioners' reports, in 1837 the legislature passed "An
act to preserve the harbor of Boston, and to prevent encroachments therein,"
establishing lines in Boston Harbor beyond which no private wharves could be
extended.[15] The harbor line laws were the first attempt to negotiate what an
upland owner could do with land between the high and low watermarks. Inevitably, litigation arose questioning the legislature's powers to limit, control, and
regulate the exercise of upland owner's rights in these lands.

In *Commonwealth v. Alger,*[16] the defendant asserted his right to extend his
wharf to a point beyond the harbor line, because it was above the low watermark
and thus within his proprietary title as granted by the colonial ordinances.[17] The
Supreme Judicial Court (SJC) rejected the defendant's argument and confirmed
the legislature's authority to regulate privately owned tidelands in the public
interest by stating: "[T]he legislature has power, by a general law affecting all
riparian proprietors on the same line of shore equally and alike, to make reasonable regulations, declaring the public right, and providing for its preservation by
reasonable restraints, and to enforce these restraints by suitable penalties."[18]

The harbor line legislation and the SJC decision in *Alger* allowed the legislature to prohibit private development on tidelands beyond the harbor line. Tidelands shoreward of that point, however, were still generally unregulated. The
harbor line legislation and *Alger* did, however, provide the legal precedent for
such regulation should the legislature decide to use it.

The Massachusetts legislature did not wait long to expand the scope of harbor regulation. In 1850, one of a series of legislatively established commissions
formed to study the effect of filling the shores or flats decided:[19] "The demand for
land is . . . an individual demand,—the demand of companies engaged in spec-
ulations; while the demand for water is a demand of the public,—a demand of
commerce,—in which the State and nation have a deep and vital interest. When
such claims come in competition, that of the public should prevail."[20]

13. *Boston Waterfront Development Corp. v. Commonwealth,* 378 Mass. 629, 638, 393 N.E.2d 356,
361 (1979).

14. 1835 Mass. Acts c. 62.

15. 1837 Mass. Acts c. 229. This law was subsequently amended several times in response to
various surveys.

16. 61 Mass. 53 (1851).

17. Id. at 56–57.

18. Id. at 95.

19. See *Report of the Commissioners in Relation to the Flats in Boston Harbor,* Sen. Doc. No. 3, at
12, 15, 16, 18 (1850).

20. Id.

Based on these studies the legislature took the first step toward gaining order and control over Boston Harbor by establishing a permanent Board of Harbor Commissioners in 1866, which was delegated the authority to pass on tideland development.[21] Under the 1866 act, every person seeking to build or fill tideland was required to notify and obtain written approval by the majority of the board.[22] Further, the Board of Harbor Commissioners was granted the authority to protect the public interest in tidelands,[23] to establish harbor lines,[24] and to prescribe plans for all tideland construction.[25]

This legislation was refined in a series of acts over the next eight years.[26] These acts further detailed the procedure required to obtain approval of tidelands projects and characterized the approval as a license. In addition, upland owners were required to compensate the state for the displacement of tidewater under such licenses.[27] From 1874 until the 1983 and 1986 amendments to the present-day tidelands licensing system under Massachusetts General Law chapter 91, the only significant developments in the Massachusetts tidelands statute pertained to the designation of the appropriate agency for the administration of the licensing laws.[28] The sole area of controversy which arose concerned the power to revoke licenses.[29] During this period, it was generally believed that licenses relied upon in good faith were the equivalent of unconditional fee simple titles. Further, the licensing legislation since 1869 had led licensees to believe that revocation of a license relied upon in good faith would be compensated.[30]

21. 1866 Mass. Acts c. 149.

22. Id. at 4.

23. Id. at 2.

24. Id. at 3.

25. Id. at 4.

26. 1869 Mass. Acts c. 432; 1872 Mass. Acts c. 236; 1874 Mass. Acts c. 284.

27. 1874 Mass. Acts c. 284.

28. For example, in 1866 the Board of Harbor Commissioners was created. 1866 Mass. Acts c. 149. Then in 1879 that board was replaced with the Board of Harbor and Land Commissioners, which assumed the powers and duties of both the Board of Harbor Commissioners and the Board of Land Commissioners. 1879 Mass. Acts c. 263. The board was then replaced with the Massachusetts Commission on Waterways and Public Lands. 1916 Mass. Acts c. 288. In 1919 the Department of Public Works and its Division of Waterways and Public Lands was established after governmental reorganization. The Division of Waterways and Public Lands assumed the duties of the commission. 1919 Mass. Acts c. 350, 111–16. In 1963 another reorganization left the jurisdiction of tidelands licensing in the hands of the Division of Waterways within the Department of Public Works. 1963 Mass. Acts c. 821. Finally, in 1974 the Division of Waterways became a division within the Department of Environmental Quality Engineering. 1974 Mass. Acts c. 806, sec. 8.

29. Legislative Research Council, *State Licensing and Control of Use of Tidelands*, House Doc. No. 2627, at 20–21 (1959).

30. This section was later codified in Mass. Gen. L. ch. 91, sec. 15. It read in part: "[S]uch authority or license shall be revocable at the discretion of the general court and shall expire in five years from its date, except as to valuable structures, fillings or enclosures actually and in good faith built or made under the authority or licensed during the term thereof." This section was subsequently amended in 1983 and 1986.

In 1941 the SJC shattered the prevailing opinion by stating that a chapter 91 license could be revoked without compensation despite the presence of valuable structures on the licensed premises.[31] This decision posed a significant risk to owners who had made improvements on their tidelands in reliance on their licenses. Further, the decision cast doubt on the ability of future tideland developers to gain adequate financing, since those who incurred expenses to improve or use licensed property did so at the risk that the license could be revoked without compensation, leaving lenders with little or no security.

In order to establish certainty of title and make land subject to a license acceptable security for the financial community, tideland developers began to request special acts from the Massachusetts legislature to declare their tideland licenses irrevocable. Between 1959 and 1969 the legislature issued forty-five so-called irrevocable licenses in tidelands areas.[32] Although the term "irrevocable license" is misleading, since a subsequent legislature can always revoke a prior enactment, developers and financial institutions gained some comfort based upon their legal advisers' confident opinion that such an irrevocable license constituted a vested property right which would require eminent domain compensation in the event the license was revoked.[33]

C. THE COURT AND LEGISLATURE REAFFIRM THE PUBLIC TRUST DOCTRINE

Aside from the irrevocability issue, public trust law remained both static and limited from the adoption of the first licensing statute until the end of the 1970s. An upland owner generally could obtain a license as long as his project did not interfere with navigation and was structurally sound.[34] The public's other reserved interests, fishing and fowling, did not enter into the decision to license a project or not.[35] However, with the SJC's 1979 decision in *Boston Waterfront Development Corp. v. Commonwealth,* the courts, the legislature and various administrative bodies began a process of refocusing on the nature of and interests protected by the public trust doctrine that led to the wholesale revision of the tidelands licensing system in an effort to rationalize and improve the commonwealth's regulation of its precious coastal heritage.

31. *Commissioners of Public Works v. Cities Services Oil Co.,* 308 Mass. 349, 366 (1941).
32. Rice, note 1 above, at 35–36.
33. Lahey, "Waterfront Development and the Public Trust Doctrine," 70 Mass. L. Rev. 55, 57 (1985). The Faneuil Hall redevelopment was executed only after such an irrevocable license was granted in 1969.
34. Lahey, note 33 above, at 55.
35. Rice, note 1 above, at 33–35.

The *Boston Waterfront* case involved a legislative grant of tidelands to the Boston Waterfront Development Corporation's (BWDC) predecessor in interest. The BWDC sought to register title to this filled and improved parcel of waterfront land, which lay seaward of the historical low watermark, claiming that it owned the property in "fee simple absolute," meaning subject to no limitations or conditions upon its use.[36] The SJC rejected the BWDC's claim, holding that the corporation had fee simple title to the property (meaning that the BWDC had the right to possess it and transfer it free of the claims of other persons), but that this title was "subject to the condition subsequent that it be used for the public purpose for which it was granted."[37] Presumably, if such land ceased to be used for such public purpose, the state could reclaim the land.

Although the *Boston Waterfront* case reasserted the public's rights in tidelands, it left a number of questions unanswered. First, the SJC did not determine what constitutes a proper public purpose.[38] Second, the court did not describe what remedy was available when public trust lands were used in a manner inconsistent with the public purpose for which the lands were granted.[39] Third, before a remedy for inconsistent use could be determined, knowledge of the original purpose was required. Until recently, however, grants of tidelands did not specify the purpose of the grant, making the test impossible. Fourth, the court never addressed the key question whether the Massachusetts legislature could ever extinguish public rights in tidelands by conveying title in fee simple absolute to private parties.

The legislature quickly stepped in to try to provide answers to these questions. In 1981 the state senate considered a bill purporting to release the commonwealth's residual rights in filled tideland within the city of Boston.[40] The purpose of the bill, as expressed in the preamble, was to provide immediate certainty "as to title to certain lands . . . bordering on or near the waters of the commonwealth, which bear clouds by implication of conditions."[41] The preamble further expressed concern that these "clouds seriously affect the value of such existing land and structures and cause serious impediments to the marketing and financing of improvements to such lands."[42] The senate asked the SJC for an advisory opinion on the bill in light of *Boston Waterfront* and previous Massachusetts case law.

36. *Boston Waterfront*, 378 Mass. at 630.
37. Id. at 649.
38. Id.
39. Id. at 654.
40. S.R. 1001 (1981).
41. Id.
42. Id.

The SJC advisory opinion, *Opinion of the Justices*,[43] reexamined and distinguished the *Boston Waterfront* decision. The court emphasized that the statute involved in the *Boston Waterfront* case did not expressly convey all of the commonwealth's interests in the tidelands to the BWDC's predecessor in interest. Furthermore, the *Boston Waterfront* case only involved submerged land lying seaward of the historic low watermark.[44] The senate bill, in contrast, specifically concerned the conveyance of *all* the commonwealth's interests in filled tidal flats.[45] The SJC concluded that the "transfer or relinquishment of all the Commonwealth's and the public's rights in the tidelands is not constitutionally beyond the power of the Legislature."[46]

With respect to tidal flats, the court concluded that once flats are lawfully filled (pursuant to either grant or license), the commonwealth can extinguish public rights in those lands, subject, however, to specific limitations. The SJC identified four criteria that grants extinguishing public rights must meet: the legislation must explicitly define the land involved; it must explicitly acknowledge the public interest being surrendered; it must expressly recognize the new use to which the land is to be put; and, most important, a "valid public purpose" must be the reason for a grant that extinguishes public rights.[47] In determining whether a valid public purpose exists, the test is "whether the expenditure confers a direct public benefit of a reasonably general character . . . as distinguished from a remote and theoretical benefit . . . and whether the aspects of private advantage . . . are reasonably incidental to carrying out a public purpose in a way which is within the discretion of the Legislature."[48]

Finally, the SJC also found it permissible to delegate to an administrative official or agency the legislative power and responsibility to regulate tidelands subject to the public trust, "provided the delegating act sets forth clearly the legislative purposes and policies and provides reasonable standards to be applied."[49]

The commonwealth's Executive Office of Environmental Affairs, particularly those administrators involved with the Massachusetts Coastal Zone Management Program established in 1978, took the opportunity created by *Boston Waterfront*

43. 383 Mass. 895 (1981). Massachusetts is one of the few states whose constitution (Amend. art. 85) authorizes the highest court to provide advisory opinions on important issues of law where requested by the governor or the legislature.

44. Id. at 901.

45. Id. at 902–3.

46. Id. at 902.

47. Id. at 905. The fact that the court only discussed the relinquishment of public rights regarding filled land suggests a fifth criterion, namely, that the public trust interests in such land be *de minimis*. Lahey, note 34 above, at 50. These requirements mirror those found in case law in other states that have held public rights could be extinguished in limited appropriate circumstances. See generally, conveyancing discussion in ch. 2, sec. F.

48. Id. (citations omitted).

49. Id. at 919–20. See discussion in ch. 2, sec. D.

and *Opinion of the Justices* to modernize the chapter 91 licensing scheme. As a result, the Massachusetts legislature amended chapter 91 in 1983 and again in 1986,[50] significantly altering the tidelands licensing scheme in place since the system was first created in 1866. These statutory amendments answered many of the questions raised by the case law discussed above and delegated the administration of the statute to the Department of Environmental Quality Engineering (since renamed the Department of Environmental Protection).

The amendments partly nullified the SJC's 1941 *Cities Services Oil* decision and provided limited financial protection in order to end the practice of obtaining irrevocable licenses by special act.[51] Chapter 91 now requires the payment of compensation when a license is revoked for any reason other than noncompliance with license conditions. Such license revocations are treated as takings of real property and thus require just compensation. The financial protection, however, is only offered prospectively to the holders of licenses issued after January 1, 1984.[52] This protection was accompanied by further restrictions on an upland owner's use of public trust land. The amendments for the first time expressly brought formerly filled flats under the statute's licensing requirements.[53] Furthermore, in keeping with *Boston Waterfront,* chapter 91 now provides that any substantial change in the use of trust land or substantial alteration of a structure thereon requires a new license, with its attendant opportunity for the state to reassess the project's consistancy with public trust principles.[54]

Most important, the amendments initiated a system for choosing among competing legitimate activities be giving priority to water-dependent over non-water-dependent uses. Water-dependent uses (generally defined as uses that require direct access to or location on the water) can be licensed without the need for a public hearing.[55] Any non-water-dependent use, on the other hand, requires a public hearing and a written determination that it serves "a proper public purpose and that said purpose shall provide a greater public benefit than that said purpose public detriment to the rights of the public in said lands."[56]

D. IMPLEMENTATION ISSUES AND PROBLEMS

Even though the 1983 and 1986 amendments' codification of the public trust doctrine into chapter 91 has significantly clarified the rights of the public and

---

50. 1983 Mass. Acts c. 589, secs. 20–27, 1986 Mass. Acts c. 348.
51. Mass. Gen. L. ch. 91, sec. 15 (1988 & Supp. 1989).
52. Id.
53. Id. at 1.
54. Mass. Gen. L. ch. 91, sec. 18 (1986).
55. Mass. Gen. L. ch. 91, sec. 18 (1988 & Supp. 1989).
56. Id. See discussion on competing uses in ch. 2, sec. B.

private parties in tidelands, the inherent complexity of balancing those rights is evidenced by the fact that final regulations implementing the amended provisions of chapter 91 were not promulgated until 1990.[57] Even as amended, chapter 91 left a number of issues open for debate at the administrative level. The following description of DEP's attempt to administer the legislative scheme while revising the chapter 91 regulations, and the discussion of several of the problems of implementation which have created the greatest controversy in the promulgation of the new regulations, may hold some lessons for other states seeking to invigorate or improve their public trust programs.

During the seven years between the 1983 amendments and the promulgation of the new regulations, the Division of Waterways of the Department of Environmental Protection met countless times with vociferous advocates for boating, development, marina, conservation, public recreation, and municipal government groups to try to reach a consensus on the implementation of the public trust doctrine through regulations. A Tidelands Advisory Committee, consisting of representatives of all these interests, met regularly with the division to comb through draft after draft of proposed regulations and explanatory reports. As a result of this extensive review, the regulations originally proposed by the division, containing numerous inflexible use and dimensional restrictions, evolved into the promulgated regulations, in which use and dimensional standards attempt to balance public and private interests while recognizing the diversity of the Massachusetts coastline. Similarly, in response to the development community's concerns over problems obtaining financing, the twenty-five-year license term first proposed was increased to thirty years, with the ability to obtain a sixty-five-year license for projects on submerged land and a ninety-nine-year license for projects on filled land.[58]

The regulations also recognize the interests and expertise of coastal communities in the management of their particular coastlines. The regulations encourage communities to develop municipal harbor plans, specifying appropriate controls on coastal development in accordance with state harbor plan guidelands. After a community adopts a state-approved harbor plan, local zoning requirements appropriate to the community are substituted for the statewide standards otherwise applied under the regulations.[59] Furthermore, the state will

---

57. Regulations implementing chapter 91 were enacted in 1978, but were not revised to address in full the 1983 and 1986 amendments until 1990. For a general overview of the 1990 regulations, see J. A. Pike and H. W. Vaughan, "Massachusetts Tidelands Laws and Regulations," 77 Mass. L. Rev. 98–111 (Sept. 1992).

58. 310 C.M.R. sec. 9.15 (1990).

59. 310 C.M.R. sec. 9.34 (1990).

generally defer to the community's judgment as to whether a particular project or public trust land serves a proper public purpose and thus should be licensed.[60]

The new regulations are not free from controversy. The regulations specify rights of the public in private tidelands that exceed those explicitly recognized in Massachusetts case law to date, including the granting of broad rights of access over tidal flats to the public for purposes beyond those to which case law has traditionally restricted such access, namely fishing, fowling, and navigation. Litigation will undoubtedly arise over the regulations and imposition of public access requirements on private developments not validated by the case law. The Division of Waterways may respond that requiring broad public access ensures that a proper public purpose is served by the licensed project. It will be up to the courts to decide whether the division has exceeded its delegated authority under chapter 91 or whether the access requirements constitute a taking without just compensation.[61] In any event, presumably the legislature could amend chapter 91 to require such public access in exchange for future licenses in the relatively unlikely event the courts rule against the agency.[62]

The division received numerous comments throughout the review period opposing regulation of formerly filled private tidelands. The case law, especially *Opinion of the Justices,* appears to support these challenges but leaves room for argument. The division has acknowledged the uncertain state of the law surrounding filled private tidelands but takes the position that it must in the first instance decide doubtful issues in the public's favor and leave it to the courts to resolve the issue. This position seems both reasonable and responsible. If the division were to exempt all formerly filled private tidelands from regulation and the courts subsequently were to find public interests do remain in such lands, many upland owners could find themselves ordered to remove or modify improvements constructed in the intervening period in reliance on the division.[63]

As a compromise, the regulations purport to exempt from chapter 91 jurisdiction any filled tidelands inland of a public way and at least 250 feet from presently flowed tidelands.[64] Again, it can be argued that this exemption amounts to a conveyance of the state's public trust interest in filled tidelands. It is unclear whether the division has the authority to create such an exemption, or whether the legislature alone can do so. Again, this issue raises delegation, as well as

60. The state does retain ultimate control over harbor plan projects through its authority to disapprove and otherwise regulate harbor plans.

61. See ch. 2, sec. H.

62. See generally ch. 2, secs. B, D, and G, and ch. 3, sec. A.

63. This situation would be similar to that created by *Cities Services* and *Boston Waterfront,* which disrupted settled private expectations.

64. 310 C.M.R. sec. 9.04 (1990).

conveyancing, questions that will almost undoubtedly be litigated if not resolved by the legislature first.[65]

The difficulties that the Division of Waterways, the public, and upland owners have had in arriving at agreement on the substantive portions of the regulations provide an excellent example of the inevitable conflict between competing bona fide public and private interests in the coastal area. The division's responses, and the remaining unresolved issues, additionally illustrate both the problems coastal managers face in implementing broad statutory pronouncements and the opportunities available to them for allocating coastal resources to maximize the reach of the public trust doctrine.

65. See discussion of legislative delegation at ch. 2, sec. D and discussion of conveyances of public trust lands at ch. 2, sec. F.

VI
Conclusion

A primary purpose of this book has been to demonstrate that the public trust doctrine has a fundamental core of precepts and organizing principles that are common among all the coastal states, notwithstanding the variations in the doctrine's scope and content. In short, the similarities among the several states with respect to the public trust doctrine are far greater than the differences.

An equally important goal of this book is to provide coastal managers, state attorneys general and other state officials, judges, environmental advocates, private property owners, title companies, and others who are interested in our valuable and limited public trust resources with a meaningful framework for: analyzing the statutory and judicial status of the public trust doctrine in their state; understanding the doctrine's reach and potential in light of the available precedents; appreciating the fact that the public trust doctrine is not self-executing but requires positive awareness and action on the part of public officials and coastal advocates; and implementing public trust principles in all regulatory decisions and actions that affect public trust lands and resources so as to protect the public interest while remaining sensitive to both the inevitable conflicts between bona fide competing interests and the legitimate expectations of private property owners.

Before beginning any analysis of the public trust doctrine or of the availability of public trust principles, coastal administrators and other advocates should carefully survey the law in their states and the precedents from other states. Having conducted a study of the law in their state, coastal managers and other advocates should use this work to identify and understand the significant issues facing them in applying or confronting the public trust doctrine. Specifically, the reader should consider the critical issues that have been addressed in this book in any analysis of the public trust doctrine:

1. The geographic reach of the public trust doctrine and procedures for mapping or identifying the boundaries

2. The categories or types of appropriate or protected uses of public trust lands and resources

3. Precedent, criteria, and procedures for choosing among legitimate competing trust uses

4. Identification of the state agencies or officials empowered to act as trustee over public trust lands

5. The procedural mechanisms available for coordination of the activities of state agencies with overlapping responsibilities regarding public trust lands

6. The procedural mechanisms available for enforcement of the public trust doctrine

7. The extent of the authority of the state or its agencies to convey public trust lands into private ownerships, and the extent to which such conveyances may operate to terminate the trust

8. The scope of the state's authority to reassert the public trust against privately owned or occupied former trust lands

9. The scope of state or agency authority to issue licenses and leases for activities on or affecting public trust lands

10. The scope of the state's authority to regulate injurious use of upland or land not traditionally defined as public trust land under the public trust doctrine or a combination of the doctrine and other legal principles, such as nuisance law

11. The vulnerability of state or agency exercises of authority under the public trust doctrine to challenges under the takings clause of the United States Constitution or the state's constitution

12. The extent to which explicit or implicit public trust principles are contained in the existing state constitution, statutes, or coastal zone programs

13. Increasing the explicit use of the public trust doctrine and public trust principles to support or justify coastal management decisions, regulations, and programs

14. Using the public trust doctrine and public trust principles to influence state or agency decision making in the most commonly occurring or controversial areas of coastal zone management

15. The scope of the state's ability to affect the use of federally owned coastal lands formerly subject to the state's public trust

16. The proper role of the federal government with respect to public trust lands under state jurisdiction

For legislators and coastal administrators, this list should be viewed as issues not just to be identified but to be addressed by careful planning and the formulation of both general policies and practical strategies. In other words, as we have advocated in this book, legislators and coastal administrators can and should take an active role in shaping the application of the public trust doctrine and public trust principles in their state, by developing strategies and mechanisms to incorporate the public trust doctrine into state constitutions, statutes, and regulatory programs, as well as to structure court cases creating favorable judicial precedents.

This analysis is particularly important in states that do not have an extensive and robust common law public trust jurisprudence, and therefore do not have a significant body of judicial precedent upon which to base decision-making processes. Absent such precedents, states and agencies do have authority, as described in this book, to inject public trust principles affirmatively into the rulemaking and adjudicatory actions taken pursuant to their enabling legislation and delegated authority over public trust lands and resources. The book should, however, be equally important for states that have already incorporated the public trust doctrine and public trust principles in much of their judicial, legislative, or regulatory framework. In such states, conducting a thorough review of the law of the state and an analysis as described in this work should lead to improved and more effective management and decision making with respect to their coastal lands and resources.

The public trust doctrine is not a magic talisman whose incantation or invocation will by itself resolve the difficult problems of our coastal areas, particularly striking the appropriate balance between public and legitimate private interests or between conflicting but valid public purposes. But when the doctrine is conscientiously analyzed and consistently applied, with judgment and sensitivity, it provides coastal managers, as well as coastal advocates, with a powerful additional tool for the vital task of allocation and management of scarce coastal resources for the good of the public at large.

Appendix

1. Is there a statute (such as the state Coastal Zone Management Act) expressly conferring public trust authority on a specific agency? Does that authority include rule-making powers? Do statutes confer public trust authority and obligations on more than one agency in the state, so that they are effectively state cotrustees?

2. If there is no explicit public trust statute, is there an existing statute or statutory program that can be or has been construed to confer public trust authority on a specific agency (or several agencies) even if the words "public trust" are not actually used? If so, under existing judicial precedent, can public trust authority impliedly possessed by an agency under such a statute or program be implemented via the promulgation of regulations incorporating public trust principles? Does the state Administrative Procedure Act apply to such an agency?

3. What are the geographic boundaries and extent of public trust authority? Do they include, in addition to coastal areas, adjacent dry sands and uplands? navigable waterways? wetlands?

4. What private uses and activities are permitted in public trust areas? Must they be, to some significant degree, water dependent before being allowed by the agency, or is any essentially public purpose sufficient? Are there any guidelines for choosing between or balancing competing legitimate public trust uses? Does the agency have authority to create such guidelines by regulation? How does the agency determine when a private project creates sufficient public benefit to be a proper purpose?

5. What is the extent of the agency's licensing/leasing authority with respect to areas covered by the public trust? For how long a term can licenses or leases for public trust uses be granted? What reasonable conditions can be attached to such

leases and licenses to protect the public interest? How much can be reasonably charged for such leases and licenses? Assuming no express statutory direction, can the revenues therefrom be used for any authorized agency purpose or only for purposes relating to the public trust?

6. What is the extent of the agency's power to enforce the public trust, for example, to prevent or limit private owners of land subject to the public trust from using their property in ways that violate or subvert the interests protected by the public trust? Have statutes or judicial decisions authorized another agency or official, or private citizens, to enforce the public trust?

7. To what extent have statutes or judicial decisions empowered or obligated the public trust agency to act affirmatively with respect to public trust lands so that the agency must take positive steps to enhance their value or increase their access, as opposed to merely reacting to prevent damage to them?

8. Has it been determined, by statute or judicial decision, whether areas covered by the public trust can be divested or conveyed to private parties; and, if so, how and by whom can such transactions be lawfully effected? If a condition of such transfer is continued use for a public purpose, how is such a condition to be monitored and enforced?

9. Is it clear, from statutes or judicial decisions, what the relationships and lines of respective authority are between the public trust agency and, on the one hand, other state agencies with some aspect of jurisdiction or power over areas covered by the public trust and, on the other hand, federal agencies with powers over tidelands such as the Army Corps of Engineers and the EPA? What mechanisms for coordination have been or should be developed? What options are available to a public trust agency when another state or a federal agency proposes or takes action inconsistent with public trust principles?

10. Are agency officials clear in their understanding of the differences between, the relative advantages and disadvantages of, and the strategies for combining their powers under the police power, the public trust doctrine, the Coastal Zone Management Act, traditional public nuisance principles, and eminent domain powers?

11. Assuming that the agency has public trust authority, should the public trust doctrine be asserted in (or attempted to be extended to) novel or controversial situations if there is a risk that the particular factual circumstances are not

favorable and could lead to an adverse judicial decision restricting the applicability of the public trust doctrine; or should the agency rather wait for the emergence of more favorable facts on which to act that will likely produce a positive judicial ruling supporting or advancing the public trust, despite the countervailing risk that agency inaction in the former situation will be argued to have created justifiable reliance and de facto precedent against the agency's later action?

Index